Physics and Philosophy

# PHYSICS
# AND PHILOSOPHY

by Sir James Jeans

ANN ARBOR PAPERBACKS
The University of Michigan Press

Third printing 1966
First edition as an Ann Arbor Paperback 1958
First published by the Cambridge University Press
Reprinted by special arrangement
All rights reserved
Published in the United States of America by
The University of Michigan Press and simultaneously
in Rexdale, Canada, by Ambassador Books Limited
Manufactured in the United States of America

# CONTENTS

# PREFACE

The aim of the present book is very simply stated; it is to discuss—and to some extent to explore—that borderland territory between physics and philosophy which used to seem so dull, but suddenly became so interesting and important through recent developments of theoretical physics.

The new interest extends far beyond the technical problems of physics and philosophy to questions which touch human life very closely, such as materialism and free-will. Thus I hope the book may interest many who are neither physicists nor philosophers by profession, and to this end I have made the discussion as simple as possible, avoiding technicalities when I could, and, when I could not, explaining them. I have also tried to arrange the book so that a reading of the first two chapters and the last shall give an intelligible view of the main argument and conclusions of the whole; many readers may prefer to read these three chapters first.

I need hardly add that my acquaintance with philosophy is simply that of an intruder, and nothing could be further from my intentions than to pose as an authority on questions of pure philosophy. If I had to choose a sub-title for my book, it might well be 'The reflections of a physicist on some of the problems of philosophy'.

I gratefully record my thanks to Sir Frederick Berryman for reading the whole of the proofs for me, to Sir Arthur Eddington for reading part (although we did not always agree), and to both, as also to Professor J. B. S. Haldane, for various criticisms and suggestions. I also thank my wife for helping me with the typing of my manuscript.

J. H. JEANS

DORKING
*July* 8, 1942

# PHYSICS AND PHILOSOPHY

## CHAPTER I

## WHAT ARE PHYSICS AND PHILOSOPHY?

Science usually advances by a succession of small steps, through a fog in which even the most keen-sighted explorer can seldom see more than a few paces ahead. Occasionally the fog lifts, an eminence is gained, and a wider stretch of territory can be surveyed—sometimes with startling results. A whole science may then seem to undergo a kaleidoscopic rearrangement, fragments of knowledge being found to fit together in a hitherto unsuspected manner. Sometimes the shock of readjustment may spread to other sciences; sometimes it may divert the whole current of human thought.

Events of this last kind are rare, but instances come readily to mind. We are likely to think first of the results of replacing the geocentric astronomy of mediaeval times by the Copernican system —man saw that his home was not the majestic fixed centre of the universe round which all else had to revolve, but one of many fragments of matter which were themselves revolving round a very ordinary one of the myriads of stars in the sky. Or we may think of the implications of the Darwinian biology—man saw that his body had not been specially designed for himself, the lord of creation, but was an adaptation and development of the bodies of animals which had preceded him on earth, and were in fact his own ancestry; all terrestrial creatures, even the meanest, proved to be his blood-relations, and if he had dominion over them it was only because he happened to have been born into the clever branch of the big family.

A third such rearrangement of ideas occurred when Newton's system of mechanics and law of gravitation gained general acceptance—men saw that the heavenly bodies were no longer to be feared or even consulted as influences in human affairs; they were only chunks of inert matter moving as they were driven by universal laws. The Newtonian scheme of things seemed further to suggest—although it was never quite able to prove—that all bodies, even the smallest, were subject to the same scheme of universal law, so that all change and motion were mechanical in their nature, the future following from the past with the inevitability of the motions of a machine. If this were so, man's imagined freedom to choose between good and evil or select his own path through life was a pitiable illusion; the ball could only go where the player sent it.

A fourth such revolution has occurred in physics in recent years. Its consequences extend far beyond physics, and in particular they affect our general view of the world in which our lives are cast—in a word, they affect philosophy. The philosophy of any period is always largely interwoven with the science of the period, so that any fundamental change in science must produce reactions in philosophy. This is especially so in the present case, where the changes in physics itself are of a distinctly philosophical hue; a direct questioning of nature by experiment has shown the philosophical background hitherto assumed by physics to have been faulty. The necessary emendations have naturally affected the scientific basis of philosophy and, through it, our approach to the philosophical problems of everyday life. Are we, for instance, automata or are we free agents capable of influencing the course of events by our volitions? Is the world material or mental in its ultimate nature? Or is it both? If so, is matter or mind the more fundamental—is mind a creation of matter or matter a creation of mind? Is the world we perceive in space and time the world of ultimate reality, or is it only a curtain veiling a deeper reality beyond?

The primary aim of the present book is to discuss the interrelation between physics and philosophy. While the discussion is in general terms, it naturally has very special reference to the changes of recent years, and their bearing on philosophical questions such as those just mentioned. But as a preliminary let us consider the general questions: What is physics and what is philosophy?

## WHAT IS PHYSICS?

Both physics and philosophy had their beginnings in those dim ages in which man was first differentiating himself from his brute ancestry, acquiring new emotional and mental characteristics which were henceforth to be his distinguishing marks. Foremost among these were an intellectual curiosity out of which philosophy has grown, and a practical curiosity which was ultimately to develop into science.

For primitive man, thrown into a world which he did not understand, soon found that his comfort, his well-being, and even his life were jeopardized by this want of understanding. Inanimate nature seemed helpful and friendly to him at times, but could become hostile when the life-giving sunshine and gentle rain gave place to the thunderbolt and whirlwind; these inspired in him the same feelings of awe and fear as the wild beasts and human foes which threatened his life. His first reaction was to project his own human motives and passions on to the inanimate objects around him; he peopled his world with spirits and demons, with gods and goddesses great and small until, as Andrew Lang has said, 'all nature was a congeries of animated personalities'. Such imaginings were not confined to cave-men and savages; even Thales of Miletus (640–546 B.C.), astronomer, geometer and philosopher, maintained that all things were 'full of gods'.

Primitive man endowed these personalities with characteristics and qualities almost as definite as those of his real friends and foes. In so doing he was not altogether wrong, for they seemed to be creatures of habit; what they had done once they were likely to do again. Even the animals understand this; they avoid a place where they have suffered pain in the past, suspecting that what hurt them once may hurt them again, and they return to a place where they have once found food, considering it a hopeful place in which to look for more. What were mere associations of ideas in the brains of animals readily became translated into natural laws in the minds of thinking men, and led to the discovery of the principle of the uniformity of nature—what has happened once will, in similar circumstances, happen again; the events of nature do not occur at random but after an unvarying pattern. Once this discovery had been made, physical science became possible. Its primary aim is to discover this pattern of events, in so far as it governs the happenings of the inanimate world.

## Positivism

The primitive stage of human development which we have just depicted is that which Auguste Comte (1798–1857) described as the stage of *fetichism*, although we now usually call it *animism*. In this stage man believed he could modify the course of events by his own volition and to his own advantage, by influencing the gods and spirits with which he had filled his world—sometimes through a policy of appeasement, as by worship and sacrifice, and sometimes through prayers, spells and incantations.

Comte says that in time this stage of animism gave place to a second stage of *metaphysics*, in which the spirits and gods of the animistic stage become depersonified, and are replaced by vaguely conceived forces, activities or essences. In this stage the world is depicted as being controlled by 'vital forces', 'chemical activities', a 'principle of gravity', and the like. These finally amalgamate into a single activity which is usually referred to as 'nature', although we still occasionally personify it and spell it with a capital N. The sequence of events has now passed beyond human control.

Comte considers that this second or metaphysical stage must in due course give place to yet a third stage—the *positive* stage. The 'forces' which expelled the spirits and gods now suffer expulsion in their turn. Nothing is left in the world but happenings for which no explanation or interpretation is offered or even attempted, and science has now for its single aim the discovery of the laws to which these happenings conform—the pattern of events.

Thus to primitive man the sun was a life-giving god—to the Greeks the horse-drawn chariot of a god—while a later and less pagan age supposed that angels had been entrusted with the task of pushing along the sun, moon and planets, and of maintaining the motion of the celestial spheres to which the more distant stars were supposed to be affixed. This animistic stage ended when the god, his horses and his chariot, the angels and their celestial spheres, were eliminated by the progress of science. To be more explicit, it ended when Copernicus, in accordance with the earlier teaching of Pythagoras, Aristarchus and others, showed how the apparent motion of the sun, moon and stars across the sky resulted from a daily rotation of the earth, while the motions of the planets through the stars could be explained by their revolutions round a fixed sun. Even when Kepler discovered the true shapes of these planetary orbits sixty years later, he still postulated a 'power' or influence to keep the planets moving; he thought they would all stop dead if a material emanation from the sun did not continually urge them on. The science of planetary movements had attained to its second stage.

Newton retained a 'force' of gravitation, but was fully conscious of the philosophical difficulties involved. When Leibniz attacked him for introducing occult qualities and miracles into his philosophy, he replied that 'to understand the motions of the planets under the influence of gravity, without knowing the cause of gravity, is as good a progress in philosophy as to understand the frame of a clock, and the dependence of the wheels upon one another, without knowing the cause of the gravity of the weight which moves the machine, is in the philosophy of clockwork'. Astronomy was beginning to move into the third stage, to which it has only recently fully attained. The astronomer of to-day makes no claim to understand why the planets move as they do; he is content to know that the pattern of events can be described very neatly and concisely by picturing planetary motions as taking place in a curved space.

Comte believed that every science must inevitably go through these three stages in turn—this is his famous 'law of the three stages'. He further claimed that the abstract sciences could be arranged in a *hierarchy*, in the order

mathematics, astronomy, physics, chemistry, biology, sociology,

in which each science is

(*a*) historically older,
(*b*) logically simpler,
(*c*) more widely applicable,

than any of the sciences which come after it on the list. Certain sciences which loom large in present-day knowledge, as for instance geology and psychology, are absent from the list and do not fit at all naturally into the hierarchy. If, however, we merge the minor sciences into the greater, the hierarchy assumes the simpler form

mathematics, physics, biology, sociology,

and now possesses all the virtues claimed for it by its author.

Comte further claimed that each science in the hierarchy is independent of all that follow it, and also must reach the final or positive stage before them. Since mathematics must have been in the positive stage from its first beginnings, the claim for physics is that it depends only on mathematics, and must be the first experimental science to attain to the positive stage. We shall investigate these claims in due course, but first let us examine the true nature of physical knowledge.

### Physical Knowledge

We each live our mental life in a prison-house from which there is no escape. It is our body; and its only communication with the outer world is through our sense-organs—eyes, ears, etc. These form windows through which we can look out on to the outer world and acquire knowledge of it. A man lacking all five senses could know nothing of this outer world, because he would have no means of contact with it; the whole content of his mind would be an expansion of what had been in it at birth.

The sense-organs of a normal man receive stimuli—rays of light, waves of sound, etc.—from the outer world, and these produce electric changes which are propagated over his nerves to his brain. Here they produce further changes, as the result of which—after a series of processes we do not in the least understand—his mind acquires *perceptions*—to use Hume's terminology—of the outer world. These give rise to *impressions* and *ideas* in turn, an impression denoting a sensation, emotion or feeling at the moment when a perception first makes its appearance in the mind, and an idea denoting what is left of an impression when its first vigour is spent, including for instance the memory of an impression or the repetition of it in a dream.

Thus the whole content of a man's mind can consist of three parts at most—a part that was in his mind at birth, a part that has entered through his sense-organs, and a part which has been developed out of these two parts by processes of reflection and ratiocination. Some have denied that the first part exists at all, holding with Hobbes (1588–1679) that 'there is no conception in a man's mind which hath not at first been begotten upon the organs of sense'—*nihil est in intellectu quod non fuerit in sensu.* Others have thought with Leibniz (1646–1716) that this should be amended by the addition of the words *nisi intellectus ipse*—there is nothing in the understanding that has not come through the senses, except the understanding itself. We shall discuss these questions more fully as the need arises.

Whenever a man increases the content of his mind he gains new knowledge, and this occurs each time a new relation is established between the worlds on the two sides of the sense-organs—the world of ideas in an individual mind, and the world of objects existing outside individual minds which is common to us all.

The study of science provides us with such new knowledge. Physics gives us exact knowledge because it is based on exact measurements. A physicist may announce, for instance, that the density of gold is 19·32, by which he means that the ratio of the weight of any piece of gold to that of a volume of water of equal size is 19·32; or that the wave-length of the line Hα in the spectrum of atomic hydrogen is 0·000065628 centimetre, by which he means that the ratio of the length of a wave of Hα light to that of a centimetre is 0·000065628, a centimetre being defined as a certain fraction of the diameter of the earth, or of the length of a specified bar of platinum, or as a certain multiple of the wave-length of a line in the spectrum of cadmium.

These statements import real knowledge into our minds, since each identifies a specific number, the idea of which is already in our minds, with the value of a ratio which has an existence in the world outside; this idea of a ratio is again something with which our minds are familiar. Thus the statements tell us something new in a language we can understand.

Each ratio expresses a relation between two things neither of which we understand separately, such as gold and water. Our minds can never step out of their prison-houses to investigate the real nature of the things—gold, water, atomic hydrogen, centimetres or wave-lengths—which inhabit that mysterious world out beyond our sense-organs. We are acquainted with such things only through the messages we receive from them through the windows of our senses, and these tell us nothing as to the essential nature of their origins. But our minds can understand and know ratios—which are pure numbers—even of quantities which are themselves incomprehensible. We can, then, acquire real knowledge of the external world of physics, but this must always consist of ratios, or, in other words, of numbers.

The raw material of every science must always be an accumulation of facts; the values of ratios of which we have just been speaking constitute the raw material of physics. But, as Poincaré remarked, an accumulation of facts is no more a science than a heap of stones is a house. When we set to work to build our house—i.e. to create a science—we must first coordinate and synthesize the accumulated piles of facts. It is then usually found that a great number of separate facts can be summed up in a much smaller number of general laws. This indeed is the most fundamental and also the most general fact disclosed by the experimental study of science— the stones fit together and combine, out of their intrinsic nature, to make a house. In brief, nature is rational. The house, being a rational structure and not a shapeless pile of stones, will show certain marked features. These express the pattern of events for which we are searching.

In physics the separate stones are numbers—the ratios just described—and the features of the house are relations between large groups of numbers. Clearly these relations will be most easily recorded and explained by embodying them in mathematical formulae, so that our scientific house will consist of a collection of mathematical formulae; in this way, and this alone, can we express the pattern of events. To take a simple illustration, the physicist finds that the spectrum of atomic hydrogen contains the line H$\alpha$

which we have already mentioned, and also a very great number of other lines which are usually designated as H$\beta$, H$\gamma$, H$\delta$, etc. The wave-lengths of these lines can be measured, and are found to be related with one another in a very simple way which can be expressed by a quite simple mathematical formula. This is typical of the way in which the particular scientific house of physics is built up; a great number of separate facts of observation are all subsumed in a single mathematical formula, and our knowledge of the physical world is expressed by a number of such formulae.

## Pictorial Representations

But now the complication intervenes that our minds do not take kindly to knowledge expressed in abstract mathematical form. Our mental faculties have come to us, through a long line of ancestry, from fishes and apes. At each stage the primary concern of our ancestors was not to understand the ultimate processes of physics, but to survive in the struggle for existence, to kill other animals without themselves being killed. They did not do this by pondering over mathematical formulae, but by adapting themselves to the hard facts of nature and the concrete problems of everyday life. Those who could not do this disappeared, while those who could survived, and have transmitted to us minds which are more suited to deal with concrete facts than with abstract concepts, with particulars rather than with universals; minds which are more at home in thinking of material objects, rest and motion, pushes, pulls and impacts, than in trying to digest symbols and formulae. The child who is beginning to learn algebra never takes kindly to $x$, $y$ and $z$; he is only satisfied when he is told that they are numbers of apples or pears or something such.

In the same way, the physicists of a generation ago could not rest content with the $x$, $y$ and $z$ which were used to describe the pattern of events, but were for ever trying to interpret them in terms of something concrete. If, they thought, there is a pattern, there must be a loom for ever weaving it. They wanted to know what this loom was, how it worked, and why it worked thus rather than otherwise. And they assumed, or at least hoped, that it would prove possible to liken its ultimate constituents to such familiar mechanical objects as occur in looms, or perhaps to billiard-balls, jellies and spinning-tops, the workings of which they thought they understood. In time they hoped to devise a model which would reproduce all the phenomena of physics, and so make it possible to predict them all.

Such a model would, they thought, in some way correspond to the reality underlying the phenomena. No one seems to have considered the situation which would arise if two different models were found, each being perfect in this respect.

Yet this situation is of some interest. If it arose, there would be no means of choosing between the two models, since each would be perfect in the only property by which it could be tested, namely the power of predicting phenomena. Neither model could, then, claim to represent reality, whence it follows that we must never associate any model with reality, since even if it accounted for all the phenomena, a second model might appear at any moment with exactly the same qualifications to represent reality.

To-day we not only have no perfect model, but we know that it is of no use to search for one—it could have no intelligible meaning for us. For we have found out that nature does not function in a way that can be made comprehensible to the human mind through models or pictures.

If we are to explain the workings of an organization or a machine in a comprehensible way, we must speak to our listeners in a language they understand, and in terms of ideas with which they are familiar—otherwise our explanation will mean nothing to them. It is no good telling a crowd of savages that the time-differential of the electric displacement is the rotation of the magnetic force multiplied by the velocity of light. In the same way, if an interpretation of the workings of nature is to mean anything to us, it must be in terms of ideas which are already in our minds—otherwise it will be incomprehensible to us, and cannot add to our knowledge. We have already seen what types of ideas can be in our minds—ideas which have been in our minds from birth, ideas which have entered our minds as perceptions, and ideas which have been developed out of these primitive ideas by processes of reflection and ratiocination.

Such ideas as originated in perceptions, and so entered our minds through one or more of the five senses, may be classified by the sense or senses through which they entered. Thus the content of a mind will consist of visual ideas, auditory ideas, tactile ideas, and so on, as well as more fundamental ideas—such as those of number and quantity—which may be inborn or may have entered through several senses, and more complex ideas resulting from combinations and aggregations of simpler ideas, such as ideas of aesthetic beauty, moral perfection, maximum happiness, checkmate or free trade. It is useless to try to understand the workings of nature except in terms of ideas belonging to one or other of these classes.

For instance, the pitch, intensity and timbre of a musical sound are auditory ideas; we can explain the functioning of an orchestra in terms of them, but only to a person who is himself possessed of auditory ideas, and not to one who has been deaf all his life. Colour and illumination are visual ideas, but we could not explain a land-scape or a portrait in such terms to a blind man, because he would have no visual ideas.

Clearly complex ideas of the kind exemplified above can give no help towards an understanding of the functioning of inanimate nature. The same is true of ideas which have entered through the senses of hearing, taste and smell—as for instance the memories of a symphony or of a good dinner. If for no other reason, none of these enter into direct relation with our perceptions of extension in space, which is one of the most fundamental of the things to be explained. We are left only with fundamental ideas such as number and quantity, and ideas which have entered our minds through the two senses of sight and touch. Of these sight provides more vivid and also more important ideas than touch—we learn more about the world by looking at it than by touching it. Besides number and quantity, our visual ideas include size or extension in space, position in space, shape and movement. Tactile ideas comprise all of these, although in a less vivid form, as well as ideas which are wholly tactile, such as hardness, pressure, impact and force. For an explanation of nature to be intelligible it must depend only on such ideas as these.

### Geometrical Explanations of Nature

Various attempts have been made to explain the workings of nature in terms of visual ideas alone, these depending mainly on the ideas of shape (geometrical figures) and motion. Three examples drawn from ancient, mediaeval and modern times respectively are:

(1) The Greek explanation that all motion tends to be circular because the circle is the perfect figure geometrically, an explanation which remained in vogue at least until the fifteenth century (p. 107, below), notwithstanding its being contrary to the facts.

(2) The system of Descartes, which tried to explain nature in terms of motion, vortices, etc. (p. 107, below). This also was contrary to the facts.

(3) Einstein's relativity theory of gravitation, which is purely geometrical in form. This, so far as is known, is in complete agreement with the facts.

We shall discuss this last theory in some detail (p. 117, below). In brief, it tells us that a moving object or a ray of light moves along a geodesic, which means that it takes the shortest route from place to place, or again, roughly speaking, that it goes as nearly in a straight line as circumstances permit. This geodesic is not in ordinary space, but in an ideal composite space of four dimensions, which results from blending space and time. This space is not only four-dimensional but is also curved; it is this curvature that prevents a geodesic being an ordinary straight line. Efforts have been made to explain the whole of electric and magnetic phenomena in a similar way, but so far without success.

It is perhaps doubtful whether such a curved four-dimensional space ought to be described as a visual idea which is already in our minds. It may be only ordinary space generalized, but if so it is generalized out of all recognition. The highly trained mathematician can visualize it partially and vaguely, others not at all. Unless we are willing to concede that the plain man has the idea of such a space in his mind, we must say that no appreciable fraction of the world has been really 'explained' in terms of visual ideas.

Even if it had, such an explanation would hardly carry any conviction of finality or completeness to our modern minds. To the

Greek mind the supposed fact that the stars or planets moved in
perfect geometrical figures provided a completely satisfying explanation of their motion—the world was a perfection waiting only
to be elucidated, and here was a bit of the elucidation. Our minds
work differently. Optimism has given place to pessimism, at least
to the extent that we no longer feel any confidence in an overruling
tendency to perfection, and if we are told that a planet moves in a
perfect circle, or in a still more perfect geodesic, we merely go on to
inquire: Why? When Giotto drew his perfect circle, his pencil was
not guided by any abstract compulsion to perfection—if it were,
we should all be able to draw perfect circles—but by the skill of his
muscles. We want to know what provides the corresponding
guidance to the planets, and this requires that the purely visual
ideas of geometrical form shall be supplemented by the addition of
tactile ideas.

### Mechanical Explanations of Nature

Explanations which introduce tactile ideas—forces, pressures
and tensions—are of course dynamical or mechanical in their
nature. It is not surprising that such explanations also should have
been attempted from Greek times on, for, after all, our hairy ancestors had to think more about muscular force than about perfect
circles or geodesics. Plato tells us that Anaxagoras claimed to be
able to explain the workings of nature as a machine. In more
recent times Newton, Huyghens and others thought that the only
possible explanations of nature were mechanical. Thus in 1690
Huyghens wrote: 'In true philosophy, the causes of all natural
phenomena are conceived in mechanical terms. We must do this,
in my opinion, or else give up all hope of ever understanding anything in physics.'

To-day the average man probably holds very similar opinions.
An explanation in any other than mechanical terms would seem incomprehensible to him, as it did to Newton and Huyghens, through
the necessary ideas—the language in which the explanation was conveyed—not being in his mind. When he wants to move an object,
he pulls or pushes it through the activity of his muscles, and cannot
imagine that Nature does not effect her movements in a similar way.

Among attempted explanations in mechanical terms, the New-tonian system of mechanics stands first. This was supplemented in due course by various mechanical representations of the electro-magnetic theories of Maxwell and Faraday (p. 120, below). All envisaged the world as a collection of particles moving under the pushes and pulls of other particles, these pushes and pulls being of the same general nature as those we exert with our muscles on the objects we touch.

We shall see later in the present book how these and other attempted mechanical explanations have all failed. Indeed the progress of science has disclosed in detail the reasons why all failed, and all must fail. Two of the simpler of these reasons may be mentioned here.

The first is provided by the theory of relativity. The essence of a mechanical explanation is that each particle of a mechanism ex-periences a real and definite push or pull. This must be objective as regards both quantity and quality, so that its measure will always be the same, whatever means of measurement are employed to measure it—just as a real object must always weigh the same whether it is weighed on a spring balance or on a weighing-beam. But the theory of relativity shows that if motions are attributed to forces, these forces will be differently estimated, as regards both quantity and quality, by observers who happen to be moving at different speeds, and furthermore that all their estimates have an equal claim to be considered right. Thus the supposed forces cannot have a real objective existence; they are seen to be mere mental con-structs which we make for ourselves in our efforts to understand the workings of nature. A simple specific example of this general argument will be found below (p. 121).

A second reason is provided by the theory of quanta. A mechan-ical explanation implies not only that the particles of the universe move in space and time, but also that their motion is governed by agencies which operate in space and time. But the quantum theory finds, as we shall see later, that the fundamental activities of nature cannot be represented as occurring in space and time; they cannot, then, be mechanical in the ordinary sense of the word.

In any case, no mechanical explanation could ever be satisfying and final; it could at best only postpone the demand for an explanation. For suppose—to imagine a simple although not very likely possibility—that it had been found that the pattern of events could be fully explained by assuming that matter consisted of hard spherical atoms, and that each of these behaved like a minute billiard-ball. At first this may look like a perfect mechanical explanation, but we soon find that it has only introduced us to a vicious circle; it first explains billiard-balls in terms of atoms, and then proceeds to explain atoms in terms of billiard-balls, so that we have not advanced a step towards a true understanding of the ultimate nature of either billiard-balls or atoms. All mechanical explanations are open to a similar criticism, since all are of the form '*A* is like *B*, and *B* is like *A*'. Nothing is gained by saying that the loom of nature works like our muscles if we cannot explain how our muscles work. We come, then, to the position that nothing but a mechanical explanation can be satisfying to our minds, and that such an explanation would be valueless if we attained it. We see that we can never understand the true nature of reality.

## *The Mathematical Description of Nature*

In these and similar ways, the progress of science has itself shown that there can be no pictorial representation of the workings of nature of a kind which would be intelligible to our limited minds. The study of physics has driven us to the positivist conception of physics. We can never understand what events are, but must limit ourselves to describing the pattern of events in mathematical terms; no other aim is possible—at least until man becomes endowed with more senses than he at present possesses. Physicists who are trying to understand nature may work in many different fields and by many different methods; one may dig, one may sow, one may reap. But the final harvest will always be a sheaf of mathematical formulae. These will never describe nature itself, but only our observations on nature. Our studies can never put us into contact with reality; we can never penetrate beyond the impressions that reality implants in our minds.

Although we can never devise a pictorial representation which shall be both true to nature and intelligible to our minds, we may still be able to make partial aspects of the truth comprehensible through pictorial representations or parables. As the whole truth does not admit of intelligible representation, every such pictorial representation or parable must fail somewhere. The physicist of the last generation was continually making pictorial representations and parables, and also making the mistake of treating the half-truths of pictorial representations and parables as literal truths. He did not see that all the concrete details of his picture—his luminiferous ether, his electric and magnetic forces, and possibly his atoms and electrons as well—were mere articles of clothing that he had himself draped over the mathematical symbols; they did not belong to the world of reality, but to the parables by which he had tried to make reality comprehensible. For instance, when observation was found to suggest that light was of the nature of waves, it became customary to describe it as undulations in a rigid homogeneous ether which filled the whole of space. The only ascertained fact in this description is contained in the one word 'undulations', and even this must be understood in the narrowest mathematical sense; all the rest is pictorial detail, introduced to help out the limitations of our minds. Kronecker is quoted as saying that in arithmetic God made the integers and man made the rest; in the same spirit we may perhaps say that in physics God made the mathematics and man made the rest.

To sum up, physics tries to discover the pattern of events which controls the phenomena we observe. But we can never know what this pattern means or how it originates; and even if some superior intelligence were to tell us, we should find the explanation unintelligible. Our studies can never put us into contact with reality, and its true meaning and nature must be for ever hidden from us.

## WHAT IS PHILOSOPHY?

Such is physics, but it is less easy to say what philosophy is. While most philosophers seem to have had their private and differing views on the question, few have been willing to venture on a definition. Hobbes (1588–1679) defined it as 'a knowledge of effects from their causes and of causes from their effects'—in other words the philosopher differs from the physicist only in that he tries to discover the pattern of events in the world at large, and not only in inanimate nature. Hegel (1770–1831) took a different view, defining philosophy as 'die denkende Betrachtung der Gegenstände', the investigation of things by thought and contemplation, again suggesting a relation—although a different one—to science, which is the investigation of things by experiment and direct inquiry. While the workshop of the scientist is his laboratory, or perhaps the open field or the star-lit sky, that of the philosopher is his own brain.

In whatever ways we define science and philosophy, their territories are contiguous; wherever science leaves off—and in many places its boundary is ill-defined—there philosophy begins. Just as there are many departments of science, so there are many departments of philosophy. Contiguous to the department of physics on the scientific side of the boundary lies the department of metaphysics on the philosophical side—that department of philosophy which lies 'beyond physics'. The boundary here is clearly defined, at least if we accept the positivist view of physics explained above. For then we must agree with Comte that the task of physics is to discover and formulate laws, while that of philosophy is to interpret and discuss. But the physicist can warn the philosopher in advance that no intelligible interpretation of the workings of nature is to be expected.

In view of this contiguity, it is not surprising that many philosophers have been physicists also. Indeed from the beginnings of recorded history down to the end of the seventeenth century—from the times of Thales, Epicurus, Heraclitus and Aristotle down to those of Descartes and Leibniz—the great names in philosophy were often great names in science as well.

It is, however, hardly possible to understand the true relation between physics and philosophy until we have glanced at some of the many forms which philosophy has assumed in the course of its long history. Without attempting anything like a sketch of the general history of philosophy (which would lie quite outside the scope of the present book), we may perhaps trace certain threads which run clearly through this history.

## Ancient Philosophy

Ancient European philosophy was almost exclusively Greek, and to the Greeks philosophy was simply what its name implies—the love of wisdom. Yet the Greek idea of wisdom was not quite the same as our own; their wisdom was based more on speculation, conjecture and contemplation, and less on firm knowledge or bedrock facts, which they had but little capacity for acquiring. In brief, it was less scientific than ours. Nevertheless it entered into some relation with science, for it comprised some real knowledge of mathematics, physics and astronomy, as well as a great mass of speculation as to cosmology, the fundamental structure of the world, and the principles governing the order of events.

But it was more especially concerned with 'the conduct of life, public and private', taking as its main topics for discussion such problems as the aim and meaning of life, the ethical principles of conduct, the most effective organization of human society, the best forms of government, education and so forth; as well as more abstract, but not entirely irrelevant, questions such as the meaning of justice, truth and beauty. In common language the philosopher was the man who could look beyond the narrow groove in which his daily work lay, and steer his way through life by availing himself of the accumulated wisdom of the race—a little knowledge mixed copiously with speculative conclusions drawn from this knowledge by contemplation, abstract reasoning and discussion.

## Mediaeval Philosophy

Then came those darker ages in which the bright light of Greek culture suffered eclipse, and European philosophy with it. During this period Christianity appeared and conquered a large part of the earth, introducing a new moral code and reshaping men's views as to the meaning and purpose of life. In so doing, it took over a large part of what had hitherto been the province of philosophy, since it provided dogmatic and professedly infallible answers to problems that had so far been topics for philosophic debate; guides to human conduct were no longer to be sought through the study of philosophy or the exercise of reason, but in the precepts of religion.

If philosophy retained any existence during this period, it was mainly through the Church trying to graft the dogmas of religion on to the older doctrines of Greek philosophy. It was studied almost exclusively by ecclesiastics, usually monks, and its language was Latin—the language of the Church, but not of any living people. Greek philosophy had been primarily concerned with problems of citizenship, ethics, and the search for the good and the beautiful; mediaeval philosophy with the subtleties and casuistries of theological doctrine. Greek philosophy had tried to advance by the exercise of reason and by controlled speculation; mediaeval philosophy by the barren methods of the syllogism and of logic-chopping. Greek philosophy had ever aimed at progress to higher things; mediaeval philosophy tried to instil an unquestioning acceptance of established authority and resignation to an unchanging order; the watchword was no longer *excelsior* but *semper eadem*.

And if science retained any existence through this period, it was a science of a useless kind, which concerned itself with, as we now know, wholly unprofitable quests such as the search for the philosopher's stone and the elixir of life, with alchemy and astrology, with magic and the black arts; its aims were wholly utilitarian and mostly unworthy.

## The Philosophy of the Renaissance

In the middle years of the fifteenth century, glimmers of a new light were seen; a dawn began to break, and the darkness of these dismal ages gradually gave place to a brighter period of intellectual and spiritual activity. For the first century at least, the interest was preponderatingly humanistic, its inspiration being drawn from classical literature. But with the coming of the seventeenth century, a new scientific interest also began to emerge, of an intellectual rather than of a utilitarian type; the foundations of modern science were being laid.

It began with astronomy. The world of mediaeval cosmology had consisted of a central earth equipped with a hell beneath and a heaven above in which God sat for ever on a throne at the point vertically above Jerusalem; the sun, moon and the star-bespangled sphere of heaven, which angels continually pushed round the earth, figured as mere adjuncts designed to secure the greater comfort of the earth's inhabitants. The writings of Copernicus, the speculations of Bruno and the observations of Galileo had shattered this old world beyond repair, and a new one was being built by the scientific astronomy of Galileo, of Kepler and, later, of Newton.

Physics soon experienced a similar change. The heathen gods and goddesses had long since passed into oblivion, so that nature could no longer be interpreted as the congeries of animated personalities who contended with one another and occasionally interfered capriciously in human affairs. Men now began to ask what it was, and how it functioned. In time it came to be interpreted as a vast machine—a network of cogs, shafts and thrust-bars, each of which could only transmit the motion it received from other parts of the mechanism and then wait for a new impulse to arrive.

This brought a beautiful simplicity into inanimate nature, but it also threatened to bring a most unwelcome simplicity into human life. For out of this view of nature there grew a philosophy of materialism, with Hobbes as its principal exponent and advocate. Its central doctrines were that the whole world could be constructed out of matter and motion; matter was the only reality; events of every kind were simply the motion of matter; man was only an animal with a material body, his thoughts and emotions alike resulting from mechanical motions of the atoms of this body.

If, then, the world of atoms worked with the inevitability of a machine, the whole race of men seemed to be reduced to cogs in the machine; they could not initiate but only transmit. Exhorting a man to be moral or useful was like exhorting a clock to keep good time; even if it had a mind, its hands would not move as its mind wished, but as the already fixed arrangement of its weight and pendulum directed. We could not choose our paths for ourselves; these were already chosen for us by the arrangement of the atoms in our bodies, and the imagined freedom of our wills was illusory.

Yet on this imagined freedom man had built his social system and his ethical code; it alone gave a meaning to his ideas of right and wrong, of purpose and moral responsibility; it formed the cornerstone of the religions in which his nobler aspirations and emotions lay crystallized; on it he had built his hopes of heaven and his fears of hell. Through the sufferings and trials of this world, he had consoled and sustained himself with the vision of the rich reward he would reap in a world to come, a reward which was to reimburse him a thousand times for the sacrifices and struggles he had so willingly made here—unless perchance, like Dante, he found his consolation in picturing the torments awaiting his enemies. But if human conduct was only a matter of the push and pull of atoms, all this became meaningless; it was in vain that he had starved his appetites, lacerated his body, and renounced all normal human pleasures; he was no more worthy of reward than the man who had wholeheartedly grasped at pleasure.

Never had a train of ideas seemed to touch human interests and everyday human life more closely; nothing could be of more tremendous import to the question of man's significance in the general scheme of things, and we might have expected that it would produce a turmoil at least comparable with those produced by the scientific findings of Copernicus and Darwin. And there were some, it is true, who showed great interest in the new doctrine. Bentley, Master of Trinity College, Cambridge, wrote that 'the taverns and coffee-houses, nay Westminster Hall and the very churches, are full of it', and added that from his own observation ninety-nine per cent of English infidels were Hobbists.

Yet the average man, who was no infidel, gave no countenance to the new doctrine—partly perhaps because he was not prepared to face its religious implications, but even more, we may conjecture, because it made no appeal to his common sense. He was perfectly clear in his mind that his will was free, no matter what abstruse arguments might be adduced to the contrary—was he not conscious of choosing freely at almost every moment of his life? Even though he might conceivably be mistaken in this, the world around him was so obviously a world of purposeful activity—men tried *and they succeeded*. The whole intricate fabric of civilized life was a standing record of achievement, not by atoms pushed and pulled by blind purposeless forces, but by resolute minds working to pre-selected ends.

Not only so, but the new doctrines of science merely restated, in rather more exact language, ideas which had long formed part of the common stock of philosophy and theology. We have already seen how Anaxagoras had explained the world as a machine in which every part moved only as directed by some other part. Seneca, again, had maintained that God 'has determined all things by an inexorable law of destiny which He has decreed and Himself obeys'. Some fifteen hundred years later, the Archbishops, Bishops and clergy of the Anglican Church, assembled in Convocation in London in the year 1562, agreed on very similar ideas which they incorporated in their Articles of Religion, and ordered to be printed in every Book of Common Prayer. After another eighty years Descartes, who certainly tried hard not to say anything that was not entirely orthodox, wrote: 'It is certain that God has fore-ordained all things', and 'The power of the will consists only in this, that we so act that we are not conscious of being determined to a particular action by any external force.'

In other words, the great machine follows its foreordained course, and we small cogs are compelled unwittingly to acquiesce in its motion—which is just about what science was beginning to say on the subject.

## Religion and Science

Although the conclusions of science accorded well enough with theological dogma on the questions of free-will and predestination, they entered into no relation at all with the teachings of pastoral theology. The preacher did not tell his flock that God had fore-ordained all things, but exhorted them to try to accomplish things of their own volition, to strive after virtue and righteousness, and in brief to attempt precisely those things which their Articles of Religion pronounced to be impossible. He did not tell them they were unable to choose, but rather that an eternity of bliss or torment depended on the choice they made.

The plain man might be content to place himself and his thoughts unreservedly in the hands of his spiritual teachers, but others saw that there was a case for investigation. It seemed to be a case for philosophy to decide and yet, if philosophy was to sit in judgment, its verdict might well seem to be a foregone conclusion. It is said that a man's philosophy is determined by his personality, or, in Fichte's words: 'Tell me of what sort a man is, and I will tell you what philosophy he will choose,' and the history of human thought supplies many confirmations of the truth of this remark. As Prof. W. K. Wright has said: 'No one in the seventeenth century but a lonely excommunicated Jew like Spinoza would have snatched at the mechanistic side of Descartes and Hobbes and given it a spiritual interpretation that could afford peace and serenity to his own tortured soul. Only enthusiastic lovers of the strenuous life like Leibniz and Fichte could have found ground for unqualified optimism in the prospect of an immortal life of unceasing activity. No one but a neurotic and selfish lover of success, with a distaste for having to work for it, such as Schopenhauer, would have seen in such a prospect the justification for a philosophy of unqualified pessimism and world renunciation. The philosophy of every great thinker is the most important part of his biography.' To which we may surely add that the biography of every great thinker is the most important part of his philosophy.

Now most of the great thinkers of this period had rather similar biographies. They lived in a highly religious age in which serious men had been educated to be, and mostly were, devout Christians. Thus most of the philosophers of the period, while ostensibly searching objectively and impartially for truth, and following the path of reason wherever this might lead, were nevertheless convinced in their own minds that their journeys could only end in a triumphant vindication of Christian doctrines, and a laying of the doubts which had been raised by science. Also, whatever their personal convictions may have been, religious feeling was so strong, and religious authority so dominant, that every writer felt himself under pressure to arrive at conclusions which conformed with the teaching of the Church; he arrived at others at his peril, as Giordano Bruno and Galileo had discovered. Further, it was an age in which consistency did not rank very high among the virtues. This is not necessarily a condemnation; it may be that we rate consistency too high to-day. Anyone whose mind is not completely petrified must find his opinions continually changing under the pressure of new experience and further consideration. And if, even at the same instant, he sees two possible solutions to a problem, no matter how inconsistent these may be with one another, there can be no reason why he should not marshal the arguments for both; he will do this in a more valuable way than two men each of whom can only see one side to the question. However this may be, even the foremost thinkers of the age we are now considering seem to have felt no embarrassment in propounding entirely inconsistent doctrines; there was even a convenient doctrine of the *twofold truth*, which proclaimed a sort of relativity of truth—a conclusion might be true in philosophy but false in theology, or vice versa.

Considerations such as these must have influenced the courses which, consciously or unconsciously, the philosophers set themselves; indeed some openly admitted their ultimate aims. For instance, in his *Critique of Pure Reason*, Kant asserted that 'The science of metaphysics has for the proper object of its inquiries only three grand ideas, GOD, FREEDOM [of the will] and IMMORTALITY, and it aims at showing that the second conception, conjoined with the first, must lead to the third as a necessary conclusion. All other subjects with which it occupies itself are merely means for the attainment and realization of these ideas.'

In the preface to the same book, Kant had explained that he had to abolish knowledge to make room for belief. 'I cannot even make the assumption—as the practical interests of morality require—of God, Freedom and Immortality, if I do not deprive speculative reason of its pretensions to transcendental insight.'

In such terms as these philosophy declared itself the handmaiden of theology.

In brief, philosophy awakened from its long mediaeval slumber to find itself confronted, among many others, with a special task. Just as the task of mediaeval philosophy had been to remove all ground for conflict between philosophy and religion, so that of the newly awakened renaissance philosophy was to avoid conflict between science and religion.

## Descartes

The foremost philosopher of this period was Descartes (1596–1650). Undeterred by his having written the sentences quoted above (p. 22), he wished above all things to maintain the freedom of the human will against the scientific considerations which seemed to be abolishing it. Apparently the crux of the whole matter, as he saw it, was the supposition that the brain consisted of ordinary matter; discredit this, and science would become harmless.

When he had written as a physiologist, he had speculated that the brain contained a fluid which he called *animal spirits*. This was neither mind nor matter, but formed a sort of intermediary between the two; mind could act on it to the extent of changing the direction, but not the amount, of its motion—for Descartes believed that the amount of motion of a material system must remain constant (p. 111)*. This fluid could in turn act on matter. To this Leibniz subsequently raised the objection that not only the total amount of motion must remain constant, but also the amount in each separate direction in space, and that any change in the directions of motion of the animal spirits would obviously change the amounts of motion in these separate directions.

When, however, Descartes wrote as a philosopher and Christian apologist, he maintained that mind was of a completely different nature from matter, and could have no contact with it. The two had entirely different functions to perform—mind to think and matter to occupy space—and they were so completely divorced that neither could affect the other to the slightest degree. In this way the will was set free, but only at the cost of creating a new problem which was to dominate philosophy for generations—if my will has no contact of any sort with the matter of my body, how can it compel this body to turn to the right or to the left as it pleases?

---

* By motion Descartes meant what we now call the momentum, $mv$. He believed that $\Sigma mv$ retained a constant value, where $\Sigma$ indicates summation over all the moving bodies. Leibniz introduced the concept of energy at a later date, describing it as force (*vis viva*, equal to $mv^2$), and found that $\Sigma mv^2$ retained a constant value. He also discovered the constancy of the momenta $\Sigma mv_x$, etc. in the separate directions in space. Descartes wanted his animal spirits to change the direction of motion while keeping $\Sigma mv$ constant. Leibniz's objection was that this would change $\Sigma mv_x$; energy did not come into the question at all.

Descartes left this problem unsolved, but we find certain of his followers—Malebranche, Geulincx, Mersenne and others, now known as the Occasionalists—solving it to their own satisfaction by supposing that the volitions of our minds are only the 'occasional' causes of the movements of our bodies, the real, ultimate, or 'efficient' cause being God. Mind and matter never interact directly, but rather run on parallel never-intersecting tracks. The good God has so arranged things that the activities of mind and matter correspond exactly to one another, and keep in such perfect step that each seems to influence the other without actually doing so. In the same way—they might have said, had they known of such things—the makers of a cinematograph film arrange that the voices and action shall correspond and synchronize through the whole length of the film; we see a soldier move smartly at the word of command, and his movement seems to be a direct consequence of the command, but actually it is the result of a pre-arranged correspondence.

### Leibniz

Leibniz (1646–1716) went further in the same direction, describing Descartes' doctrine of the distinctness of mind and matter as 'the ante-room of truth, but only the ante-room'.

Giordano Bruno had already supposed the world to consist of a number of ultimate indivisible units which he called 'monads'; these were at the same time spiritual and material in their nature. Every human being and every living thing was such a monad. The monads were all distinct and different, and could not be resolved into anything simpler.

Leibniz also supposed the world to consist of a great number of simple units, which he too described as monads—whether he borrowed the name from Bruno is not known. These monads, he says, are the true atoms of the universe, the ultimate constituents of everything, and they possess neither shape nor size nor divisibility. Now, as Plato had argued in the *Phaedo*, dissolution and decay appertain only to complex, and above all to divisible, structures. Thus their very simplicity shields the monads from dissolution and decay, so that they are necessarily eternal and immortal. Each man's soul is a single monad, and his body a collection of monads of various kinds. All substances are of the nature of force*, and consist of individual centres of force, which must thus be monads, and 'in imitation of the notion which we have of souls' must contain something of the nature of feeling and appetite. These monads, then, are more or less spiritual in their nature. The lowest monads of all, Leibniz writes, resemble animals in a swoon, higher monads have clearer perceptions and are endowed with memory, while God is the highest monad of all. Since all monads are spiritual in their nature, matter can have no real existence, and must come from seeing monads in a confused way.

The monads have no windows on to the outer world through which anything could come in or go out, so that each lives its utterly secluded life, uninfluenced by its fellow-monads. Its changes are determined only by its own internal state; it can come into existence only through a creative act of God, and can go out of existence only through annihilation by God. Yet God, the supreme monad, keeps all the other monads in step on a series of parallel tracks.

Leibniz calls this the *System of pre-established Harmony*. 'Under this system', he wrote, 'bodies act as though there were no souls, and souls act as though there were no bodies, and both act as though each influenced the other.'

Leibniz explained this further by comparing the soul and body (or mind and matter as we should now say) to two clocks which always show the same time, a comparison which the Occasionalists had used before him. There are, he says, three ways in which two clocks can be made always to show the same time. One—and here he refers to the experiments of Huyghens—is by putting them in close physical contact, so that each clock transmits its vibrations to the other, and the two clocks advance in unison; this is the solution of ordinary philosophy, but must, Leibniz thinks, be rejected because we cannot imagine anything being transmitted between mind and matter. The second way is to have a clockmaker continually putting the clocks in agreement; this also Leibniz rejects because it requires the incessant intervention of a *deus ex machina* 'for a natural and ordinary thing'. The third and only other way, says

* By 'force' Leibniz here means energy or *vis viva*.

Leibniz, is to construct the two clocks so perfectly at the outset that they will agree through all time.

This last is the way of the system of pre-established Harmony. In the beginning God created mind and matter in such a way that each can follow its own laws, and yet the two move in the same perfect agreement as would prevail 'if God were for ever putting in his hand to set them right'.

To use Leibniz' own illustration, we, with our puny abilities, can make an alarm clock and set it to sound an alarm at any hour we require. Obviously, then, so great a craftsman as God could make Caesar's body and pre-arrange its atoms so that it should go to the Senate House at such or such an hour on the Ides of March, should utter such and such words, and so on. The same great Craftsman could also create the soul of Caesar in such a way that it should experience certain emotions in a pre-arranged order and at pre-arranged moments of time, and could, if He so wished, plan that these should exactly correspond to, and synchronize with, Caesar's bodily movements. According to Leibniz, He had so wished.

The wheel had now come round full circle. In his eagerness to establish the freedom of the will, Descartes had divided the universe into two ingredients, mind and matter, which could not interact; this raised the problem of how mind and matter could keep in step without interacting. Leibniz, trying to explain this, had to suppose that neither had more freedom than a machine which, having once been set in motion, was compelled to execute a pre-destined series of mechanical movements. In this way, every mind became an automaton, which is precisely the conclusion that Descartes had been trying to escape, and one that Leibniz would presumably have liked to avoid if he could.

### Kant

So the question stood when Kant brought his mind to it. He saw that the circle of arguments of Descartes and Leibniz could lead nowhere except to the very conclusion that both were, like himself, eager to avoid. He was as much concerned as his predecessors to establish the freedom of the will, but he had a clearer conception of

the difficulties in the way. 'As the complete and unbroken connection of phenomena is an unalterable law of nature,' he wrote,
'freedom is impossible—on the supposition that phenomena are
absolutely real. Hence those philosophers who adhere to the
common opinion on this subject can never succeed in reconciling
the ideas of nature and freedom.'

By 'the common opinion' Kant meant what would now be
described as Naïve Realism, or Common-sense Realism. This rejects all metaphysical subtleties, and maintains that the phenomena
we observe correspond fairly closely to the realities of the world
outside us; when we think we see a brick at some point of space,
there really is something 'there', which is much like what we
imagine a brick to be. Thus the world is just about what it seems to
be, consisting simply of the particles and objects which are found,
by observation and experiment, to obey a causal law. If, says Kant,
this is all there is to the world, then obviously the will cannot be
free.

On the other hand, many philosophers have found it difficult to
accept the hypothesis that an object is just about what it appears to
be, and so is like the mental picture it produces in our minds. For
an object and a mental picture are of entirely different natures—a
brick and the mental picture of a brick can at best no more resemble
one another than an orchestra and a symphony. In any case, there
is no compelling reason why phenomena—the mental visions that a
mind constructs out of electric currents in a brain—should resemble the objects that produced these currents in the first instance.
If I touch a live wire, I may see stars, but the stars I see will not in
the least resemble the dynamo which produced the current in the
wire I touched. In this instance, the current produces a vision in
my mind which differs utterly from the object which created the
current. May it not be the same with all the phenomena of nature?

When we perceive an object, we perceive at most a few of its
qualities. Having perceived these few qualities, we frequently
jump to the conclusion that the object belongs to some familiar
class of object possessing these qualities. We see a kittenish patch
of colour behaving in a kittenish way, and conclude that we are
seeing a kitten. But our identification may be wrong; the little

creature may be a skunk.  Again, when a tiny meteor smaller than a pea is falling through the air, it will send the same electric currents to our brains as will a giant star millions of times larger than the sun and millions of times more distant.  Primitive man jumped to the conclusion that the tiny meteor was really a star, and we still describe it as a shooting-star.  This and innumerable other instances show that two objects may differ widely in their intrinsic natures and yet produce similar, and even identical, phenomena. And as the two objects of such a pair cannot both be like their mental images, there is no longer any sufficient reason for thinking that either of them must be.

Thus we can no longer hold that objects in general are pretty much like their mental images.  The images need not resemble the objects in which they originate, and our perception of the outer world may consist only of *representations* which are constructed by our minds out of the activities flowing into our brains, and bear little or no resemblance to the realities outside.  They may be like the code signals which the signalman sends over the wires to say what kind of train is coming next; these bear no resemblance to the train.  Or, as Boltzmann suggests, they may be merely symbols which are related to the objects as letters are to sounds, or as notes are to musical tones.

Kant, holding that phenomena are only representations, argues that they must originate in something other than phenomena, so that even though the phenomena may be bound to other phenomena by causal laws, their origins need not be.  If we limit our attention to the phenomena, our observations suggest that causality governs everything, but if we could make contact with the reality underlying the phenomena we might see that this is not so.

A few pages later, he explains that his remarks were not intended to prove the *actual existence of freedom*, or even to demonstrate the possibilities of freedom; 'that nature and freedom are at least *not opposed*—this was the only thing in our power to prove, and the question which it was our task to solve'.

Still, it seems difficult to accept this as providing even a *possible* solution to the problem of human free-will.  The average man is not interested in the origins underlying phenomena; the freedom of which he wants to assure himself, and instinctively believes himself to possess, is a freedom to control, or at least to influence, the phenomena, or, according to Kant, the representations. Imagine two men who are similar down to the last atoms of their bodies, placed in environments which again are similar down to the last atoms. If free-will is to be explained in the way Kant suggests, we can imagine one exercising his freedom and deciding on a saintly life, while the other may decide at the same moment that he is more hedonistically inclined. Up to the moment of making these choices, the phenomena have been the same for both, so that if causality prevails in the world of phenomena, as Kant supposes, the subsequent phenomena must also be the same for both; the two men must mutter the same prayers and drink similar drinks—with similar results. So far as the phenomena go, their two lives will be identical, and to their acquaintances will be indistinguishable. It follows that the men can have no moral responsibility for their actions—only at most for their intentions and desires. Clearly this is not what the plain man means by freedom of the will, and neither is it what Kant wanted to establish. But the question is no longer of more than academic interest, since, as we shall soon see, science now finds that even the phenomena are not governed by causal laws.

On other questions besides that of human free-will at which we have just glanced it was obvious that the methods of science could lead only to the conclusions of science; if philosophy was to reach other conclusions she must employ other methods. Furthermore, if she wished her conclusions to take precedence over those of science, she must be able to claim that her methods were in some way more trustworthy than the methods of science. This led to a critical examination of the methods by which scientific knowledge was obtained, and to an intensive study of certain problems of what is now called epistemology—the science of knowledge. This will form the subject of our next chapter.

# CHAPTER II

## HOW DO WE KNOW?

### (DESCARTES TO KANT; EDDINGTON)

#### THE SOURCES OF KNOWLEDGE

We have already noticed how knowledge is gained by establishing relations between an inner process of understanding in our private minds and the facts of that public outer world which is common to us all. As Plato pointed out, the use of a common language is based on the supposition that such relations can be established by all of us.

In the period we have been considering, science claimed only one source of knowledge of the facts and objects of the outer world, namely the impressions they make on the mind through the medium of the senses. Yet the untrustworthiness of the senses had been one of the commonplaces of philosophy from Greek times on, and if the same facts and objects of the outer world made different impressions on different minds, where did science stand? If we trusted to individual sense-impressions, we could never get beyond the position described by Protagoras (*c.* 481–411 B.C.): 'What seems to me is so to me, what seems to you is so to you'; each individual would become his own final arbiter of truth, and there could be no body of objective knowledge. Six centuries before Christ, in the earliest days of Greek philosophy, Thales of Miletus had urged the importance of gaining a substratum of facts, independent of the judgment of individuals, on which a body of objective knowledge could be built.

These difficulties are non-existent to the modern physicist, who can trust his instruments to give absolutely objective and unbiased information, but they loomed large when there were no instruments beyond the unaided human senses. To avoid them, Plato argued in the *Theaetetus* (*c.* 368 B.C.), we must distinguish between what the mind perceives through the senses and what it apprehends of itself by thinking. Such concepts as number and quantity, sameness and difference, likeness and unlikeness, good and bad, right and wrong do not enter our minds through our senses, but reside permanently in our minds. And as concepts such as these provide the formal element in all true knowledge, it follows that this does not come from our sensations, but rather from the judgments our minds pass on our sensations.

Plato elaborated this into an argument that the human mind is equipped from birth with a set of *forms* or *ideas* which exist in it independently of the objects of the outer world. These latter serve as a sort of raw material for the impress of the forms, so that each object becomes a sort of meeting-place for a number of forms. A red square brick, for instance, is a lump of this raw material stamped with the impresses of the forms of redness, squareness and brickiness. When we declare that a particular object is a red square brick, we mean that in our judgment this particular piece of matter fits into these three forms. We may of course be mistaken; seen in a different light, the object may appear of some colour other than red, measured against a set-square it may prove to be far from square, and hit with a trowel it may prove not to be a brick at all.

On such grounds as these, Plato maintained that we have sure and certain knowledge only of the forms and their relations; our knowledge of the objects of the outer world consists at best of fleeting impressions and shifting opinions. In the matter of reality and certainty, the ideas which reside permanently in our minds, namely the forms, may claim precedence over ideas put there temporarily by objects we perceive with our senses: it is in this world of permanent ideas which exists outside space and time, the world *sub specie aeternitatis*, that truth alone can dwell.

This train of thought retained existence of a sort through the dark ages of philosophy; it figured prominently, although in a modified form, in the philosophies of St Thomas and the scholastics, and finally reappeared, still further modified, in the philosophy of Descartes.

The ideas of Plato, the forms, had been ideas of qualities or properties; he supposed that these were inborn in our minds, as though for instance they were memories carried over from a previous existence. The ideas of Descartes, on the other hand, were ideas of facts, or propositions as we should now call them. He thought they were *innate* in a sense rather different from that of Plato; the mind was not born with these ideas inside it, but with a predisposition to acquire them as soon as it came into contact with the world. 'I called them innate in the same sense in which we say that generosity is innate in certain families and certain diseases such as gout or gravel in others—not that the infants of those families labour under those diseases in the womb of the mother, but because they are born with a certain disposition or faculty of contracting them.'

Leibniz subsequently challenged this, arguing that all ideas are innate in this sense, but that they only mature into actual thought when they have been developed by the growth of knowledge. The mind at birth is not a clean sheet of paper, but rather an unworked block of marble, in which there is already a latent structure of veins; this will to some extent determine the form the marble will assume when the sculptor chisels it into shape.

Others differed still more widely from Descartes, and in the period we now have under consideration we find the philosophers divided, broadly speaking, into two camps—the *rationalists*, who maintained that the highest truth resides in our own minds and so is to be discovered by reason, and the *empiricists*, who thought that truth resides outside our minds and so is only to be discovered by observation and experiment on the world outside.

## *The Rationalists*

The rationalists, headed by Descartes, argued that all knowledge obtained through direct observation of nature was suspect because it comes through the senses, and such knowledge can notoriously be deceptive, as all kinds of hallucinations and dreams show. Descartes added that even knowledge obtained by mathematical proof may be deceptive—first, because mathematicians have often been wrong, and second, because we can never be certain that an omnipotent God may not have decreed that we should be deceived even in the things we think we know best. In this way the rationalists discredited, even if they did not dispose of, practically the whole of scientific knowledge—it came from tainted sources. They proposed replacing it by the store of knowledge which, as they believed, was to be derived from pure contemplation.

Descartes claimed that his innate ideas, representing knowledge which came from 'the clear vision of the intellect', must necessarily be true. The fact that he could clearly and distinctly conceive something in his mind—as, for instance, the existence of God—was for him a sufficient proof of its truth. Others claimed that, inborn in the human mind, there are a number of ready-made *principles* or *faculties* by the recognition and skilful use of which it must be possible to discover truths about the universe, just as Euclid was able to discover geometrical truths from a few axioms, the truth of which was obvious. Kant went so far as to claim that it ought to be possible in this way to construct a 'pure science of nature', which should be independent of all experience of the world, and therefore uncontaminated by the errors and illusions of observation. A very similar claim has again been put forward in recent years by Eddington (p. 72, below).

Kant attempted a reasoned discussion of this question in his famous *Critique of Pure Reason*. He reminds us of Plato when he says that a phenomenon, or object of perception, contains both substance and form; the substance produces the effect in the mind of the percipient, while the form enables us to allocate the phenomenon to a wider class. The substance of a phenomenon comes to us as the result of an experience of the world, or, in Kant's terminology, *a posteriori*; but the form, which is already in our minds lying in wait for the substance, comes to us *a priori*—i.e., previously to, and independently of, all actual experience of the world.

Relations between *a priori* concepts which are such that they can be known without any appeal to experience will constitute a body of knowledge 'altogether independent of experience, and even of all sensuous impressions'. Such knowledge Kant described as *a priori* knowledge, in contradistinction to empirical or *a posteriori* knowledge, which has its sources in experience. *A priori* knowledge, then, came direct from heaven through the gates of ivory, and so was in every way superior to knowledge discovered through experiment, by observation, or even (according to Descartes) by mathematical demonstration, all of which came only through the gates of horn. *A priori* knowledge was necessarily applicable to every possible experience, whereas empirical knowledge, which was known only as the result of limited experience or observation, could make no such claims.

Also *a priori* knowledge was applicable to every possible universe, and not only to this one—for we can distinguish this universe from other possible universes only by observation, and once we do this our knowledge ceases to be *a priori*. Thus in claiming *a priori* knowledge, we claim to know enough of the ultimate nature of things to be able to say what kinds of universe a Creator could have created, and what kinds He could not have created. Kant's claim that a 'pure science of nature' is possible in principle involves just this claim. Like every other claim to *a priori* knowledge, it not only denies the omnipotence of God, but also claims to have detailed knowledge of His limitations. It is a high claim to make for the human intellect.

### The Empiricists

In opposition to this, the empiricists held that in general knowledge comes from experience alone, so that the only way to discover the facts about the universe is to go out into the world and search for them. Most empiricists were nevertheless willing to concede that certain truths could be known by intuition or through demonstrations based on intuitions.

Locke and Hume, the two most prominent of the empiricists, were in agreement that the truths of pure mathematics could be known in this way, as also are most modern philosophers, as for instance Whitehead and Russell. But J. S. Mill held the opposite view, maintaining that the laws of arithmetic embodied generalizations derived from observations of actual objects, while geometry dealt merely with idealizations of objects of experience—we could not imagine a mathematical point, line, or triangle unless we had first made the acquaintance of their imperfect representations in the outer world. Locke thought that not only the truths of pure mathematics but also the existences of God and ourselves and the truths of morality ought to be admitted to the class of intuitive truths.

The whole question is obviously largely one of words. As regards the truths of morality, for instance, the question at issue is whether God could have made a world in which a different morality would have been 'true'. And surely the answer depends at least as much on what we mean by morality and truth as on what we know about morality and truth.

In general, however, the empiricists held firmly to the principle that knowledge *about the outer world* must come from the outer world, and so can be acquired only by observation and experiment. As this is precisely the method of science, it might have been expected that those philosophers who were also scientists, or were of a scientific turn of mind, would be found in the camp of the empiricists, while those of a mystical or religious turn of mind would be found among the rationalists.

Actually almost the exact opposite was the case. I suppose the four most distinguished advocates of rationalism were (in chronological order) Descartes (1596–1650), Spinoza (1632–1677), Leibniz (1646–1716) and Kant (1724–1804). Two of these four names are among the very greatest in mathematics. Descartes was not only the father of modern philosophy but also of modern mathematics, being, amongst other things, the inventor of analytical geometry, while Leibniz shares with Newton the honour of having created the differential calculus, and incidentally anticipated Einstein by maintaining that space and time consist only of relations, in opposition to the Newtonian view that they are absolute.

Kant can make no claims comparable with these, and yet we should remember that astronomy and physics had interested him more than philosophy in his earlier years; according to Helmholtz, he only turned from science to philosophy, at the age of thirty-one, because there were no facilities for scientific research in his University of Königsberg. And he gave scientific lectures regularly to the end of his academic career, and wrote on a variety of scientific subjects, such as earthquakes, lunar mountains and the possibility of changes in the revolution of the earth. Most of his scientific work has long been forgotten, but he was the first to suggest the true nature of the external galaxies—clusters of myriads of stars— and he has the not inconsiderable distinction of having propounded one of the first theories of the evolution of the solar system. Besides introducing these evolutionary ideas into astronomy, he was one of the earliest of biological evolutionists. In his *Anthropology* he declares in favour of all animals being descended from a common ancestor, although he does not include humanity in this statement— possibly because of its dangerous religious implications. Still, he suggests that man must have changed fundamentally in the course of time, adding that in some future natural revolution orang-outangs might acquire not only human form, but also the organs of speech and the use of intelligence. He once wrote that he was 'thinking many things, with the clearest conviction and to his great satisfaction, which he would never have the courage to say'. Prof. Paneth has suggested that one of these things may well have been that what could happen to orang-outangs and chim-panzees in the future might also *have already happened* in the past. Engraved on his tomb at Königsberg are words from the end of his *Critique of Practical Reason*—'Two things fill the mind with ever new and increasing admiration and awe, the oftener and more steadily we reflect on them; the starry heavens above and the moral law within.' The order is significant.

Spinoza can advance no claims to scientific distinction, although his thought is obviously often guided by mathematical and scientific knowledge.

Against this, none of the more prominent of the empiricists— Francis Bacon (1561–1626), Locke (1632–1704), Berkeley (1685–1753) and Hume (1711–1776)—had any special scientific attain-ments; Berkeley wrote an 'Essay towards a new theory of Vision', but its scientific value is not great.

The reason for this rather strange division of forces may have been in part that those who understood science best were also most acutely conscious of its anti-religious implications. But the true line of demarcation between the two schools of thought was geo-graphical. The Continentals, with their love of abstract ideas, can claim all the rationalists, while the British, with their love of practical investigation, claim the empiricists, the four just men-tioned being English, English, Irish and Scottish respectively.

## A Priori *Knowledge*

The debate as to whether genuine *a priori* knowledge exists need hardly concern us; the question which matters for our present discussion is not whether such knowledge exists, but the much simpler question of whether, if it exists, it is important. To this it seems possible to give a negative answer without appealing to anything more recondite than the well-known principle that the proof of the pudding is in the eating. Of course we must ourselves be judge and jury, since it is an obvious impossibility for a man who does not claim infallibility to convince one who does that he is wrong. But even if I cannot persuade my cook that her puddings are bad, I can still dismiss her from my service.

The main reason which seems to call for an adverse judgment on alleged *a priori* knowledge is that it has often been proved false by the subsequent advances of science.

As examples of the kind of knowledge of which the truth was claimed to be obvious *a priori* may be taken:

'The same thing cannot at once be and not be.'

'Nothing cannot be the efficient cause of anything.'

'The liberty of our will is self-evident.'

'Everything that happens is predetermined by causes according to fixed law.'

Descartes gives the first three of these, describing the last of them as 'a truth which must be reckoned as among the first and most common notions which are born with us'. With any reasonable use of language, it is obviously in contradiction with the fourth, which is taken from Kant, so that *a priori* knowledge begins to discredit itself by its contradictions even before the evidence of science has been called.

Nothing would be gained by trying to analyse these statements in detail, but one general remark at once suggests itself. It is surely improbable, on principle, that these or any similar statements can express absolute truths when stated without qualification in the crude bald forms permitted by common language. Such words as *thing*, *cause*, *liberty* and *predetermined* mean nothing definite until they have been defined. If we are free to supply our own definitions, we shall probably be able to find a sense in which all the propositions will be true, and a sense in which all will be untrue; or we may be able to find a group of cases in which they are true and a group in which they are untrue. Thus they do not present universal truths so much as topics for debate, the question at issue being the limits or conditions within which each is true. Stated in the uncompromising terms permitted by common language, the propositions merely prejudge questions on which philosophy has broken its teeth through the ages.

Other pieces of alleged *a priori* knowledge were of a more scientific kind, and these are of more interest to our present discussion. We may take two examples from Descartes:

(*a*) the sum of the three angles of a triangle is 180°,

(*b*) divisibility is comprised in the nature of substance, or of an extended thing,

and three from Kant:

(*c*) space has three dimensions,

(*d*) between two points there can be only one straight line,

(*e*) in all changes of phenomena, substance is permanent, and the quantity thereof in nature can be neither increased nor diminished.

Kant describes (*c*) and (*d*) as principles 'which are generated in the mind entirely *a priori*', and (*e*) as a piece of knowledge which 'deserves to stand at the head of the pure and entirely *a priori* laws of nature'.

As soon as we try to discuss these propositions in the light of modern science, we again feel the need of precise definitions of the terms used. Thus (*a*) and (*d*), which are geometrical in their nature, are true in the kind of space which is defined by the so-called 'axioms' of Euclid—Euclidean space, as it is usually called—but not in the curved space in which the planets are now usually pictured as moving. Did then Descartes and Kant intend their propositions to refer to Euclidean space, or to this possibly more real curved space? The answer is almost certainly that they were thinking of Euclidean space. In Descartes' time no other kind of space was contemplated. In Kant's time, other kinds were under consideration, but Kant held that Euclidean geometry was 'true' in a sense in which other geometries were not, although admitting that he could not prove this—because the axioms of Euclid could be denied without any inconsistency or contradiction. Thus we can see now, although Descartes and Kant could not, that their supposed *a priori* knowledge cannot claim to be applicable to any objective space of the outer world, but only to private worlds of their own. In so far as they thought that their *a priori* knowledge applied to the real world, they were more wrong than right.

Kant's proposition (*c*)—that space has three dimensions—is in a different class; it is hard to see how it can claim to be *a priori* knowledge. For every mathematician knows that it is just as easy, as an abstract exercise, to imagine a space of one, two or four dimensions as one of three. If, then, a new-born baby knows that the space of the outer world has three dimensions, this must be because he has already been peeping at the outer world, or has otherwise made its acquaintance; his knowledge is empirical and not *a priori*.

It is much the same with the two remaining propositions, which are of a more physical nature. In (*b*) Descartes tells us that divisibility is a property of substance or of an extended thing, but fails to tell us what he means by *substance* or *thing*. Actually of course divisibility is a property of an elephant or a sandstorm, but not of a photon or an electron; but Descartes does not give any definition of a *thing* which will include elephants but exclude electrons. In (*e*) Kant tells us that substance is permanent, but fails to define substance. He does, however, say that his statement is tautological, which seems to imply that he would define substance as that which is permanent, in which case the statement tells us something about Kant's use of words, but still nothing about the objective world. Since Kant's time physicists have found that substantial electrons and other material particles may dissolve into, and also be created out of, insubstantial radiation. Even if these phenomena had not been observed, we now know that there is, in principle, no permanence in substance; it is mere bottled energy, and possesses no more inherent permanence than bottled beer, although it is of course true that under the physical conditions prevailing on our particular planet, matter may be regarded as very approximately permanent.

### The Three Worlds of Modern Science

It is a natural transition from this to a reflection of a very general kind, which proves to be of the utmost importance for our discussion of the bearings of science on philosophy. The human race first became acquainted with the properties of matter in the special forms they assume under the physical conditions prevailing on our planet. In the same way, the laws of nature first became known to our race in the restricted form of laws applicable to the behaviour of objects comparable in size with human bodies, the reason of course being that only objects of these sizes could be studied without elaborate instrumental aid. In such studies time was usually measured in seconds or minutes and length in inches or yards, while nothing ever moved much faster than a galloping horse.

But with the instrumental aid now at its disposal, science can study phenomena in which times are measured, maybe in fractions of a millionth of a millionth of a second, maybe in thousands of millions of years; the lengths involved may be small fractions of a millionth of a millionth of an inch, or they may be millions of millions of miles, while the objects concerned may move at a millionth part of a snail's pace or at a million times the speed of an aeroplane.

Surveying these immense ranges as a whole, we find that ordinary human activities occupy a fairly central position in the scheme of the universe; the world of man lies just about half-way between the world of the electron and the world of the nebulae. It also occupies only an excessively minute fraction of the whole range between electrons and nebulae. The smallest piece of matter we can feel, see or handle without instrumental aid still contains millions of millions of millions of atoms and electrons, while even the smallest of the planets stands in about the same relation to the largest piece of matter we can move with our unaided bodies.

Elaborate studies made with instrumental aid have shown that the phenomena of the world of the electron do not in any way form a replica on a minute scale of the phenomena of the man-sized world, and neither are these latter a replica on a minute scale of the phenomena of the world of the nebulae. As we leave the man-sized world behind us, and proceed either towards the infinitely great in one direction or towards the infinitely small in the other, the laws of nature seem at first sight to change, not only in detail but in their whole essence.

More careful scrutiny discloses that the apparent change is illusory; actually the same laws prevail throughout the range, but different features of these laws become of preponderating impor-tance in different parts of the range. A soap-bubble obeys precisely the same law of gravitation as a cannon-ball, and also precisely the same law of air-resistance. Clearly, then, we can combine these two laws into a single law which must govern the motion of soap-bubble and cannon-ball equally. But if we let the two objects fall together from the leaning tower of Pisa, their motion will seem to be governed by entirely different laws. The reason is that gravita-tion is all important for the cannon-ball, while air-resistance is all important for the soap-bubble.

In the same way, all objects are governed by the universal laws of physics, but one aspect of these laws is all important for the electron, another for man-sized objects, and yet a third for the movements of the nebulae. These three departments of the uni-versal scheme of law are so different that we are justified in thinking of them as constituting three distinct and separate sets of laws with a different pattern of events in each.

This is a fact of tremendous importance to philosophy as a whole. Its immediate importance to our present discussion is that it opens up two new worlds in which to test the alleged *a priori* knowledge of the rationalists. If this knowledge is found to be true in the two new worlds, the question of whether it is genuine *a priori* know-ledge must still remain unanswered. If, however, it is found to be untrue in either or both of the new worlds, its claim to be genuine *a priori* knowledge is obviously discredited—the a priorists have told us that the Creator could not make a world in such and such a way; we study the world of the electron or nebula and find that He has done so already. Thus the alleged *a priori* knowledge can only be empirical knowledge of the man-sized world.

Now when the actual intuitions of the rationalists are tested in these two new worlds, we find that those which were of a scientific nature are frequently not true for the two new worlds which science has just opened for us; they are only true for the man-sized world which was familiar to the rationalists because it did not need elaborate instrumental aid for its exploration. For instance, three of the examples of *a priori* knowledge just given ought, as a preliminary step towards the truth, to be amended to read:

'The sum of the three angles of a triangle is 180°, so long as the triangle is not of astronomical size.'

'Divisibility is comprised in the nature of substance, so long as the object in question is not of the smallness dealt with in atomic physics.'

'Substance is permanent, so long as we experiment only to the degree of accuracy possible for eighteenth century physics.'

No philosopher seems to have had an inkling, either *a priori* or otherwise, of the need for these or any similar reservations until modern physics arrived to point it out. The plain fact seems to be that when a rationalist, guided by his experience of the world, but subject to the scientific limitations of his day, could only imagine things being one way, he confidently announced that they were that way and had to be that way, describing his knowledge as *a priori*. Now that recent scientific investigations and discussion have opened up new worlds to the imagination, we can think soberly of possibilities that would have seemed sheer absurdities to Descartes and Kant. Not only can we imagine them, but we know that many of them find their counterparts in the actual world, and tell us that the supposed *a priori* knowledge of the rationalists was erroneous. Kant tells us that there are two infallible tests for true *a priori* knowledge—necessity and strict universality. The supposed scientific knowledge of the a priorists fails conspicuously under both tests, and this failure of their scientific intuitions naturally discredits their non-scientific intuitions. But knowledge of a mathematical kind requires further investigation.

## Mathematical Knowledge

While philosophers may have differed as to the possibility of obtaining *a priori* knowledge about the world of physics, they have been in very general agreement—apart from Descartes (p. 34) and J. S. Mill (p. 36)—that abstract knowledge of a mathematical kind could be obtained through purely mental processes, without any appeal to experience of the world, so that such knowledge can be truly *a priori*. They would have claimed this knowledge to be true in all possible worlds; it would be a knowledge of facts which it was beyond the powers of the Creator to vary. Thus it could tell us nothing about the properties of our particular world, as distinguished from those of other possible worlds which might have been created.

We have cited three instances of supposed *a priori* knowledge of this kind, all three being geometrical in their nature, but the progress of science has shown that they all three fail to qualify as true knowledge of the physical world.

Now that science is actively concerned with non-Euclidean geometries, philosophers have become chary of finding examples of *a priori* knowledge in geometry, and are more inclined to look to arithmetic or simple algebra. The proposition that two and two make four is frequently cited in this connection, although its precise content is seldom stated, so that we feel that the first need is for definitions and explanations. The simple question is: Could God have made a world in which two and two did not make four? and however much or little we may claim to know about the Creator, it is obvious that, before we can discuss this, we must know what the two and two are which form the subject of the proposition. Are they things which exist in reality or in our minds? Are they numbers or objects? And in the latter event, what kind of objects?

If the two and two refer to mere numbers, then the proposition is concerned with simple counting, and its content would seem to be a definition of the term four. We count two and then another two, and this brings us to a number to which we must give some sort of a name. The proposition tells us to call it four, although we might

equally call it something else, such as quatre or vier; as indeed
many people do. Clearly there can be no question of *a priori*
knowledge here.

Obviously then, the proposition must be interpreted as referring
to real physical objects. It tells us that if we take two objects of any,
but the same, kind, and add to them two more objects of still the
same kind, we shall then have a collection of four objects in all—
not that we shall have taken four in all, for this would bring us
back to mere counting, but shall have four objects under our
observation as the result of doing something other than counting.
The child is shown that when two apples are placed in juxtaposition
with two other apples the result is a collection of four apples; he
sees that the same is true of fingers or counters or pennies, and then
jumps to the conclusion that it is true of everything we can imagine,
as for instance bananas or sea-serpents or unicorns. The knowledge
about the apples or fingers is admittedly empirical, but this merely
serves to pull a trigger; what is claimed as *a priori* knowledge is that
we may generalize from apples and fingers to sea-serpents and
unicorns.

If this is the true content of the proposition, does it not merely
provide another instance of incomplete or ill-considered knowledge
being labelled *a priori*? For the generalization (which is the essence
of the proposition) proves to be permissible for some classes of
objects and for some circumstances, but only for some. It is
impossible to say whether it is true in any particular case without
detailed knowledge of the case, and such knowledge from its nature
can never be *a priori*. We cannot say what two sea-serpents and two
sea-serpents make until we know what a sea-serpent is, and this
cannot be *a priori* knowledge. A sea-serpent is often said to be a
cloud of birds; do then two sea-serpents when placed in juxtaposition with two more make four sea-serpents, or do they make
one big sea-serpent, or perchance two or three? And what about
two raindrops meeting two more on the window-pane? If two
negatives make a positive, while two positives also make a positive,
what results from adding two negatives to two more? Clearly the
proposition is applicable only to objects which retain their identity
through the process of physical addition, and we cannot know

*a priori* whether any particular class of objects possesses this property or not. Of recent years mathematicians have studied algebras in which two and two make numbers other than four, perhaps two or one or even zero; such algebras do not of course apply to mere numbers, but to operations, processes or events. Before we can assert that two objects plus two objects make four objects, we must find a definition of *object* that will exclude such things, and clearly this cannot be inborn in us as *a priori* knowledge.

Kant did not discuss the proposition that $2+2=4$, but the proposition that $7+5=12$. He described this as a *synthetical a priori* proposition (p. 49), meaning that a special addition with fingers was needed to pull the trigger in his mind, and suggest the truth of the general proposition. But he does not define 12, or specify the 5 and 7, other than fingers, to which the proposition is supposed to apply.

A better example would perhaps have been the proposition that $5 \times 7 = 7 \times 5$, for this at least does not require a definition of 12, or even of 5 and 7, since it is equally true if 5 and 7 are replaced by undefined numbers or numerical quantities $p$ and $q$. The proposition then states that the product $pq$ is equal to the product $qp$; in other words, when we multiply $p$ and $q$ together, the order in which we take them is a matter of indifference. This is obviously so if $p$ and $q$ denote pure numbers, but before we can assent to the general proposition, $p$ and $q$ must be defined with some care. Mathematicians now employ algebras, which they describe as non-commutative, in which $pq$ is not the same thing as $qp$; these are found to be specially applicable to the sub-atomic world. In most of the problems which arise for discussion in the man-sized world, $p$ and $q$ have such meanings that $pq$ is equal to $qp$, but in the world of the electron this is not so. We may conjecture that a denizen of the world of the electron might vigorously challenge the general proposition that $pq=qp$, insisting that it was true only under very special conditions (p. 157, below).

Thus a large part of our mathematical knowledge proves on examination to be more empirical, at least in its application, than is evident at first sight, or than appears to have been suspected by the a priorists. We may say that a general proposition, such as that $2+2=4$ can be true in either of two ways—either *a posteriori* or *a priori*. It is not true for objects in the outer world unless these conform to certain conditions. These conditions cannot even be stated, still less applied, without some knowledge of the outer world, so that when the proposition is applied to real objects, it obviously represents *a posteriori* knowledge; we first test whether the proposition is true for the class of objects under consideration, and the proposition then merely gives back to us the knowledge we have previously put into it. But the proposition can also be applied to classes of objects we imagine in our minds in such a way that they satisfy the conditions necessary for the proposition to be true. When used in this way, the proposition contains pure *a priori* knowledge, but it can never tell us anything about the outer world— only about the imaginings of our own minds.

For instance the proposition $2+2=4$ as applied to apples is *a posteriori* because we call on our experience of the world to assure us that apples retain their individual identities through the process of adding. But as applied to unicorns, it is *a priori* because the unicorn is a creature of our imagination, which we imagine to retain its identity through the process of adding.

We see that when mathematical propositions are applied to objects in the *a posteriori* manner, they can supply no knowledge about the outer world beyond that we have previously put into them, while when they are applied in the *a priori* manner, they can give us no knowledge at all about the outer world—*ex nihilo nihil fit*.

There is, nevertheless, a wide range of abstract mathematical knowledge which can be derived by purely mental processes, without introducing any knowledge of the world outside. The clearest instance of such knowledge is to be found in the properties of pure numbers or numerical quantities, as expressed in arithmetic or ordinary algebra, but we must notice that even here we have to assume that numbers and measurable quantities exist. For example, we can show by purely mental processes, and without calling on our experience of the outer world at all, that if $a$ is a pure number, then $(a+1) \times (a-1)$ is always less than $a^2$—for example, $8 \times 6$ is less than $7^2$. In the same way, it can be discovered that 8, 9 and 10 are composite numbers (i.e. numbers obtained by multiplying smaller numbers together), while 7 and 11 are primes (i.e. non-composite numbers).

Such facts involve no knowledge or experience about the particular world in which we live (unless we regard the existence of measurable quantity as an empirical fact), but, in so far as they have to do with worlds at all, are true of all worlds which could be either built or imagined. In whatever way this or any other world is constructed, 7 must be a prime, and, just because of this, the primeness of 7 can never tell us anything about the special structure of our particular world; no bridge can be built between the two. The same is true of all the discoveries of the pure mathematician; they are universal in the sense that they would be true in any world, and so cannot tell us anything about the special properties of this particular world.

Indeed any knowledge which is truly *a priori* must, as Kant says, be universal, and so can tell us nothing about our particular world. Let us imagine a totally uneducated man being told he was going to be sent to Procyon. He would not know whether Procyon was a prison or a gin-palace, an island or a star. But he would know just as much about Procyon as our unaided *a priori* knowledge can tell us about the universe we live in, and if he tried to construct a 'pure science of Procyon', his efforts would be no more futile or misguided than those of Kant to construct a 'pure science of nature'. In such ways we see that there can be only one possible source of knowledge as to the special properties of our own world, namely experiment and observation; there is only one method of acquiring such knowledge, namely the method of science.

## Synthetic Knowledge

As the admission of this obvious truth would have undermined Kant's whole position, he made two attempts to evade it; they are quite distinct, although he does not seem to have realized this.

In the first he claimed to be in possession of a special kind of *a priori* knowledge—*synthetic a priori* knowledge as he called it— which conveyed knowledge about our particular world.

In the second he claimed in effect that our physical knowledge is not knowledge about the world, but about the workings of our own minds—not knowledge of the world we perceive, but of our mode of perception of the world. Let us consider these two attempts at escape in turn.

We have already had an example of Kant's *synthetic a priori* knowledge in the proposition that $7 + 5 = 12$. A more characteristic instance is the proposition that 'all bodies are heavy'. In discussing this, Kant first cites the proposition that 'all bodies are extended' as a typical piece of *a priori* knowledge which was, in his judgment, obvious apart from all experience of the world. He then says that, after encountering extended bodies in the actual world, we find that they are heavy as well as extended. Adding this new fact to his previous knowledge he arrives at the proposition that 'all bodies are heavy'.

He considers that all the propositions of arithmetic, and many of the principles of physics, are of the *synthetic a priori* type. As instances he selects the conservation of matter (pp. 40, 41) and Newton's third law of motion (p. 108), expressing them in the words 'In all changes of the material world, the quantity of matter remains unchanged' and 'In all communications of motion, action and reaction must always be equal'.

Science can of course have nothing favourable to say about this. On Kant's own admission, he only knows of heaviness through observing it in the actual world, and this immediately removes it from the category of *a priori* knowledge—*synthetic a priori* is seen to be merely a new, and question-begging, name for *a posteriori*. In the instance just quoted, Kant's claim is in effect to know of the existence of gravitation, but if he could know of this, why did he not also know of electric attractions and repulsions? Would he have known *a priori* that two objects similarly electrified would not attract but repel one another?

In such ways Kant persuaded himself that this supposed *a priori* knowledge provided definite and certain information about the actual universe. Claims of this kind at once raise questions such as the following:

(1) If *a priori* knowledge does not come from our experience of the world, whence does it come? The rationalists claimed to have

*a priori* knowledge that everything must have a cause; what, then, is the cause of *a priori* knowledge itself?

(2) If *a priori* knowledge does not come from our knowledge of the world, how can it tell us anything about the world? How does it happen that, when we step out into the world, we find this world conforming to our *a priori* knowledge? If Kant or Eddington succeeded in constructing a whole universe out of such knowledge, on what grounds would he expect the actual universe to conform to his predictions?

## KANT'S THEORY OF KNOWLEDGE

Kant saw very vividly the difficulties presented by this and similar questions, and this led him to fall back on his second line of defence, developing a set of ideas as to the precise meaning of which philosophers themselves are not altogether in agreement. Indeed there is every justification for wondering whether Kant altogether understood them himself. Sixteen years after the publication of the *Critique of Pure Reason*, Kant's doctrines were making a great turmoil in Germany; university professors were being forbidden to lecture on them and one at least was forced to resign for venturing to disagree with him. This moment was chosen for asking Kant to say which of his commentators had best grasped his meaning. In reply, Kant indicated a certain Schultze, the author of an elementary explanation which seems to have over-elaborated the easier parts of Kant's philosophy at painful length, while dismissing the more difficult parts in a few words which were demonstrably wrong. Thus the problem of discovering what Kant had been trying to say remained unsolved, as it still is to-day. James Ward tells us that no fewer than six different formulations of Kant's philosophy were current in the years 1865 to 1878.

Although no one can say precisely what Kant meant to convey, I hope what follows will be found to express an average view as to his meaning, in so far as it affects the problems before us.

To the first of the two questions stated above—If *a priori* knowledge does not come from our experience of the world, whence does it come?—Kant's answer seems to be that *a priori* knowledge comes from the inherent constitution of the human mind. Just as the human body is built in a certain way, with two eyes and two ears and other specific organs which perform specific functions, so the human mind is built in a certain way, with specific faculties which perform specific functions. It is to these faculties that we must look for the sources of our *a priori* knowledge. They sift out the sense-data with which our senses continually deluge our minds, allowing some to slip through unheeded while retaining others.

Out of such as are retained, the mind creates its own picture of the external world. As a result of the sifting action of the mind, certain laws and regularities emerge to which all our perceptions conform. If we run a miscellaneous collection of potatoes over a sieve of one-inch mesh, we know that any potato-pattern left on the sieve will conform to at least one law—every one of its ingredients will be more than one inch in diameter. This law is not obeyed by potatoes in general, nor by the miscellaneous collection of potatoes that were passed over the sieve; it is a law thrust on to the potatoes by the selective action of the sieve, and expresses a property of the sieve rather than of potatoes. Kant suggests that those laws of nature which we know (as he thought) *a priori* are thrust on to the perceived world by a selective action of the human mind, which thus acts as a lawgiver to nature; *a priori* knowledge merely specifies the conditions to which phenomena must conform if they are to be perceived.

Possible modes of selection can perhaps be illustrated by two simple analogies. Light is a blend of constituents of different wave-lengths. If we pass the light through a spectroscope, the different constituents are separated out, and we observe a spectrum of colours ranging from red, through orange, yellow, green to blue and violet—the colours of the rainbow. Outside the limits of this spectrum all looks dark, yet if a thermometer is placed in the dark region beyond the red, the mercury begins to rise, showing that beyond the reddest of visible radiation there is an invisible radiation; it is in fact the infra-red heat-radiation. Beyond the violet at the other end of the spectrum there is another region in which our eyes can see nothing, but in which certain salts phosphoresce, showing that here too there is radiation which is invisible to our eyes. This is the ultra-violet radiation; out beyond this we come to X-radiation, and further still to the $\gamma$-radiation emitted by radio-active substances.

Our instruments reveal a continuous spectrum of rays ranging from long radio waves to short $\gamma$-rays, the wave-lengths of these extremes standing in a ratio of about twenty thousand million million to one. By contrast, the extremes of radiation that our eyes can see have wave-lengths standing in a ratio of only about two to one. Thus of the whole range of radiation known to us through our instruments, only one part in ten thousand million million is perceptible to our eyes—an infinitesimal fraction of the whole.

The restriction of our vision to so minute a part of the whole spectrum acts as a sieve to our perceptions. All sorts of radiation fall on the retina, but this is sensitive only to a small part of what it receives; it forwards such radiation, and only such, to the mind for its attention. The mind might draw the inference that all radiation lies between the red and the violet. On Kant's view this would correspond to the *a priori* knowledge claimed by the rationalists, and it may be noticed that, in so far as the analogy is sound, the only inference to be drawn is that *a priori* knowledge is wholly untrust-worthy.

It is the same with sound. Our ears are sensitive only to sounds the pitch of which lies within about ten octaves, out of the infinite range which can occur in nature. If we took the data provided by our unaided sense-organs at their face-value, we might claim to know that all sounds lay within a range of ten octaves.

Such is the way in which the physical sieves of our sense-organs work. A simple analogy may explain how our mental sieves may work. The night sky exhibits a confused mass of stars which might be sorted into constellations in many ways. The Greeks, with their minds accustomed to run on legend and romance, sorted the stars out into figures of heroes and their accompanying animals; the more prosaic Chinese saw the same groups of stars as quite commonplace animals. But there are also stars in the southern sky which the Greeks had never been able to see, because their travels were confined to the northern hemisphere. When the navigators of a later age explored the southern seas, and first saw these stars, they did not see them as groups of new heroes and animals. The age of such fancies had passed, and the explorers left it to their prosaic astronomers to group the new stars in the forms of triangles, clocks, telescopes, and so on; they chose these because their practical minds were accustomed to thinking of such things. The division of the stars into constellations tells us very little about the stars, but a great deal about the minds of the earliest civilizations and of the mediaeval astronomers.

Kant thinks that it is in such ways as these that our minds sort out the phenomena of nature. The outer world provides us with a confused mass of impressions which our minds might sort out in many ways. They choose one particular way because they are constituted in one particular way; other types of mind might choose other ways. The laws we deduce from our *a priori* knowledge or reasoning merely represent habits of thought embedded in our own minds. These habits of thought form blinkers, restricting the free vision of our minds. But the mind, not recognizing its own limitations, proceeds to attribute these limitations to nature itself. Thus, in Kant's own words, 'reason only perceives that which it produces after its own design', 'objects conform to the nature of our faculty of perception' and 'we know *a priori* of things only what we ourselves put into them'.

Kant described this as his Copernican revolution. When no further progress seemed possible to an astronomy which supposed that the sun revolved round the astronomer, Copernicus cleared up the situation by supposing that the astronomer revolved round the sun. Kant thought that he had removed the difficulties of *a priori* knowledge in a similar way—if our minds conformed to the phenomena they perceived, our knowledge could not be *a priori*; we must therefore (so Kant thought) make the phenomena conform to our minds.

If this were the true significance of *a priori* knowledge, it would of course tell very little about nature—only something about our own minds. Our knowledge would not be of the structure of the universe without, but of the structure of our minds within. Here, then, we have the answer to our second question—If *a priori* knowledge does not come from our knowledge of the world, how can it tell us anything about the world? The answer is that it cannot; it can only tell us about the structure of our own minds.

All this throws a vivid light on the different methods of science and philosophy. Kant proposed in effect that we should base our knowledge of things on something that 'we ourselves put into them'; the scientist is anxious to eradicate just this something, knowing that it is not knowledge of the outer world at all.

The 'sieves' which Kant attributed to the human mind are fourteen in number. First of all come two which he describes as 'Forms of Perception'—these are merely space and time. Then come twelve others, which could well be described as 'Forms of Understanding', although Kant preferred to describe them as 'Pure Conceptions of the Understanding' or 'Categories', this latter term being borrowed from Aristotle.

We want ultimately to bring Kant's views on space and time into relation with present-day science; for this reason we may conveniently proceed at once to discuss space and time in rather general terms.

## SPACE AND TIME

As present-day science knows, the words *space* and *time* admit of many interpretations. Four distinct meanings may be discerned for each, those for *space* being approximately as follows:

*Conceptual space* is primarily the space of abstract geometry. It has no existence of any kind except in the mind of the man who is creating it by thinking of it, and he may make it Euclidean or non-Euclidean, three-dimensional or multi-dimensional as he pleases. It goes out of existence when its creator stops thinking about it—unless of course he perpetuates it in a text-book.

*Perceptual space* is primarily the space of a conscious being who is experiencing or recording sensations. We feel an object, and our sense of touch suggests to us that it is of a certain shape and size; we see a collection of objects, and our vision suggests to us that these objects stand in certain relations to one another. We find that we can reconcile these and all other suggestions of our senses by imagining all objects arranged in a threefold ordered aggregate which we then call space. This is perceptual space, created for himself by a man experiencing sensations, and it goes out of existence as soon as his sensations cease. For a one-eyed man, or one viewing objects so remote that his binocular vision conveys no idea of distance, perceptual space is two-dimensional—at least so long as no sense other than seeing is involved. Thus the ancients located the fixed stars on the two-dimensional surface of a sphere. As soon as near objects are viewed by a normal man, so that binocular vision is employed, or as soon as objects are seen to move one behind another, or as soon as senses other than seeing are employed, a third dimension of perceptual space instantly springs into being.

*Physical space* is the space of physics and astronomy. Conceptual space and perceptual space are both private spaces, the one being private to a thinker, and the other to a percipient. Science finds, however, that the pattern of events in the outer world is consistent with, and can be explained by, the supposition that material objects are permanently located in, and move about in, a public space which is the same for all observers, apart from a complication introduced by the theory of relativity to which we shall return later (p. 63). Disregarding this complication for the moment, we may say that this public space is physical space.

*Absolute space* is the particular type of physical space which Newton introduced to form the basis of his system of mechanics (p. 108, below), and remained in general scientific use throughout the period between Newton and Einstein. When we say that a train has moved 10 miles nearer to King's Cross, we mean that it has moved a distance of 10 miles along the pair of rails along which it is running to King's Cross, as, for instance, from milestone 105 to milestone 95. In the same interval of time, the earth—carrying this pair of rails with it—may have carried the train 100 miles to the east by its daily rotation around its axis, and may have moved 10,000 miles in its yearly orbit round the sun, while the sun, dragging the earth along with it, may have moved 100,000 miles nearer to the nearest star and 1,000,000 miles farther away from a distant nebula. All these motions are equally real and equally true, but all are *relative* only to some other moving body.

The sequence might go on indefinitely, but Newton imagined

that it did not. He thought that the remotest parts of the universe were occupied by vast masses which might provide fixed points of reference from which to measure motion, while themselves providing standards of absolute rest, although he qualified this by remarking 'it may be there is no body really at rest, to which the places and motions of others can be referred'. At a later period, space was supposed to be filled with a jelly-like ether, and this again was thought to provide a standard of absolute rest until it was abolished by the coming of the theory of relativity. Assuming that such standards existed, Newton described the space in which measurements were made from them as *absolute space;* this, he said, 'in its own nature and without regard to anything external, always remains similar and unmoveable'. He contrasted it with perceptual space, which he described as *relative space*—'some moveable dimension or measure of absolute space which our senses determine'.

In a precisely similar way we may discern four distinct meanings for time; there are a conceptual time, a perceptual time, a physical time and an absolute time.

*Conceptual time* is the time of theoretical dynamics, and of all abstract attempts to study change and motion. Like conceptual space it exists only in the mind of a thinker. He usually makes it one-dimensional, but not always. Dirac, for instance, found it convenient to measure time by a $q$-number, which amounts to supposing that time has as many dimensions as we please to assign to it.

*Perceptual time* records the flow of time for any single percipient. Thus it is related to the consciousness of a particular individual, and goes out of existence as soon as this individual loses consciousness. Experience shows that the acts of perception of every percipient lie on a single linear series—in other words, they come *one after another*. Thus perceptual time is one-dimensional.

*Physical time* is the time of the active world of physics and astronomy. Like physical space it is public, in contrast with conceptual and perceptual time, which are private. Again disregarding complications introduced by the theory of relativity, science finds that the pattern of events is consistent with the supposition that all events can be arranged uniquely in a single linear sequence, the position on this sequence determining the time. This still permits of an infinite number of ways of measuring the time, so that a convention must be introduced as to how the actual measure is made. We agree to select some motion which repeats itself regularly, such as that of the earth in its orbit, to form a 'clock', and let each repetition of this motion count as a unit of time—in this case a year. But as this unit is too large for most practical purposes, other regularly repeating motions must be found, such as the oscillations of a pendulum or the vibrations of a crystal, which repeat many times in a year, and these provide the units needed for ordinary life and for scientific investigations in which time is involved.

*Absolute time* is the counterpart of absolute space. We have just seen how a 'clock' can be devised to give a consistent measure of time at any one point of space. The problem of synchronizing clocks in different parts of space is a different problem, to which we shall return later. If light travelled with infinite speed, it would be as simple, in principle, to synchronize distant clocks as it is to set our watches by Big Ben. Newton, disregarding the finite speed of travel of light, assumed that this could be done, and that a universal time 'flows equably, and without regard to anything external' throughout the universe. This we describe as absolute time.

## What are Space and Time?

There can be no serious difficulty in understanding the meaning of conceptual and perceptual space and time, for they are our own creations. They exist in our individual consciousnesses, and go out of existence when these consciousnesses cease to function. But a variety of views can be held, and have been held, as to the true significance of physical space and time.

Science has usually adopted a *realist* view of the world of nature, assuming that our perceptions originate in a stratum of *real* objects—stars, bricks, atoms, etc.—which exist outside, and independently of, our minds. If our minds go out of existence or cease to function, the stars, bricks and atoms continue to exist, and are still capable of producing perceptions in other minds. On this view, space and time have just as real existences as these material objects; they existed before mind appeared in the world, and will continue to exist after all mind has gone.

But philosophy has pointed out that other views are possible. We can have no knowledge except self-knowledge; what is in our minds we know, but what is outside we can only conjecture. And our conjectures may be erroneous. The *mentalist* or *idealist* philosophies suppose that there is no stratum outside all mind having an existence of its own in the way the realists suppose; consciousness is fundamental in the world, and the supposed real objects which produce our perceptions are creations either of our own or of some other minds (p. 196). There is no reason to attribute a higher degree of reality to space and time than to the objects we locate in space and time, so that these also become mental creations. Conceptual and perceptual space and time are now as real as anything there is, while physical space and time become attempted mental generalizations of these realities—in strong contrast to the realist view which makes physical space and time the realities, while conceptual and perceptual space and time are mere reflections of, and abstractions from, these realities.

The first modern to discuss the nature of space and time was Nicholas of Cusa (1401–1464). He held that space and time are products of the mind, and so are inferior in reality to the mind which has created them. In contrast with this purely philosophical conclusion, Giordano Bruno (1548–1600), discussing space and time in their astronomical aspects, argued that the words 'above' and 'below', 'at rest' and 'in motion' become meaningless in the world of eternally revolving suns and planets which know of no fixed centre. Thus all motion is relative—as Einstein subsequently convinced the world—and absolute space and time must be figments of the imagination. Leibniz (1646–1716) held very similar opinions, believing that space and time exist only relative to objects and not in their own right; space is merely the arrangement of things that co-exist, and time the arrangement of things that succeed one another. All these thinkers, then, reduced space and time simply to conceptual and perceptual space and time; physical space and time had no real existence, and absolute space and time did not come into the picture at all.

In opposition to them all came Isaac Newton (1642–1727), tacitly assuming that space and time were no mere dependents on consciousness but existed in their own rights, and introducing the hypothesis that absolute measures of space and time were possible, at least in principle.

## Kant's Discussion of Space and Time

Kant began his discussion of space and time by asking the questions: What are space and time? Are they real existences? Or are they merely relations between things? And in this case, would these relations belong to the things even though the things should never be perceived, or do they belong only to things when these are perceived—i.e. are they contributions of the perceiving mind?

He made no distinction between the different kinds of space and time that we have mentioned, identifying them all with perceptual space and time. His general view was that space has no real existence of its own, but is supplied by our minds as a framework for the arrangements of objects, so that it is only from the human point of view that we can speak of space, the extension of objects and so forth. 'Space is not a conception which has been derived from outward experience; it is a necessary representation *a priori*, which serves for the foundation of all external perceptions.' Time again is not an empirical conception and has no real existence of its own, but whereas space serves for the representations of external perceptions, time serves for the representations of internal perceptions—'the perception of self and of our internal states'.

Kant tries to justify these views in the discussion of his 'first antinomy'. By an antinomy Kant means a pair of more or less contradictory assertions, each of which seems to be proved by disproving the other. In his own words, we originate a conflict of assertions 'not for the purpose of finally deciding in favour of either side, but to discover whether the object of the struggle is not a mere illusion, which each strives in vain to reach but which would be no gain even when reached'. 'Perhaps after [the combatants] have wearied more than injured each other, they will discover the nothingness of their cause of quarrel, and part good friends.' A new set of ideas which reconciles the combatants is described as a solution of the antinomy. It may or may not be true; its truth is established only if it can be shown to provide a *unique* solution of the antinomy, but not otherwise—a point which Kant overlooks.

Kant's first antinomy consists in brief of the assertions that it is impossible to imagine either that

(*a*) the world had a beginning in time, and is also limited in space,

or (*b*) the world had no beginning in time, and has no limits in space.

The reasons he gives for dismissing both alternatives as absurd seem entirely unconvincing to a modern scientific mind. There is, of course, no justification for tying up an infinity of space with an infinity of time in the way that Kant does. Mathematicians have investigated the properties of universes in which space is finite but time infinite, and no logical inconsistency has so far been detected in the concept. It is, however, quite simple to discuss time and space separately.

In opposition to alternative (*b*), Kant argues that any quantity must be regarded as the synthesis of a succession of separate unit quantities. For example, a mile must be regarded as the length of 1760 yardsticks put end to end. If, then, the quantity is infinite, the synthesis can never be completed; this, he says, is the true definition of infinity. Hence 'it follows, without possibility of mistake, that an eternity of actual successive states up to a given (the present) moment cannot have elapsed, and that the world must therefore have a beginning.'

In this argument the words 'can never be completed' are obviously ambiguous. We want to know who or what can never complete them, why he should want to, and whether he wants to complete them in his imagination or in some sort of reality; until this information is given us, the argument is simply a meaningless collection of words.

Apart from this, the argument fails because a quantity can be regarded in other ways than as a succession of units. Must we always think of a mile as 1760 yards? Why this rather than as a succession of eight furlongs? And why either rather than as just one mile? Yet as soon as we concede the last possibility, the bottom drops out of Kant's argument, since we need only increase our unit *pari passu* with the length of space or time to be measured. Even though our finite lives may be too short to imagine eternity as a succession of hours or years, we can still think of it as one eternity.

In opposition to (*a*), on the other side of the antinomy, Kant argues that if the world had a beginning in time, there must have been a previous void time in which there was no world. But there can be no reason for anything beginning in a void time, since 'no part of any such time contains a distinctive condition of being, in preference to that of non-being'. Thus the world cannot have had a beginning.

This argument fails through assuming that time would necessarily go further back than the beginning of the world. This has not been the usual view of philosophy. Plato, for instance, said that time and the heavens came into being at the same instant; Augustine wrote, 'Non in tempore, sed cum tempore, finxit Deus mundum', while Kant himself tells us that time does not subsist of itself, but is 'the form of the internal state, that is, of the perceptions of ourselves and of our state'. But if time is in ourselves, and we in the world, then time must be in the world, and it is a *petitio principii* to argue as though the world were in time.

After adducing arguments of somewhat similar type for space, Kant proposes the solution that space and time have no real existences, but are only forms of human perception. As they are, then, creations only of the human mind, we are free to imagine alternative (*a*) at one moment and (*b*) at the next, if we so wish; the two assertions of the antinomy become no more contradictory than the uses of a Mercator Projection and a stereographic projection in map-making, and we are free to use whichever serves our purpose best. But even if Kant's arguments were sound, we should be under no obligation to accept his proposed 'solution' of the antinomy, for he does not even attempt to prove that it is the only possible solution.

Three general reflections on the problem of space and time will perhaps not be out of place here in view of their bearing on Kant's doctrines of space and time.

## The Finite Speed of Light

Light takes time to travel through space, a fact which does not appear to have been known to Kant, although it had been discovered by the Danish astronomer Roemer as far back as 1675. Jupiter has a number of moons which circle round it with the same regularity as that of the moon round the earth. When the precise period of revolution of any moon of Jupiter has been found, it would seem to be a simple matter to draw up a time-table of its future movements. Roemer made such a time-table, and discovered that the moons did not keep to it; they seemed to get late and run behind their scheduled time whenever Jupiter was at more than its average distance from the earth, and to be ahead of time when Jupiter was at less than its average distance. He found, however, that all the observations could be explained by supposing that light travelled through space at a uniform finite speed; the apparent irregularities of Jupiter were then explained by the variations in the time which light took to travel from the planet to the earth. The truth of this explanation was established beyond all doubt when Bradley discovered the phenomenon of aberration in 1725.

This shows that space and time are not totally independent of one another as Kant and many others seem to have imagined; on the contrary there must be a fairly intimate connection between them.

## The Space-Time Unity

The theory of relativity has revealed the nature of this connection. Newton supposed that all objects could be located in his absolute space, and that all events, wherever they occurred, could be assigned positions uniquely and objectively on an ever-flowing stream of absolute time. These assumptions provided him with an approximation which was good enough for his purpose, and fitted in with the scientific knowledge of the seventeenth century. Subsequent investigation has shown that they are inadequate to explain the passage of light and the behaviour of objects moving at a speed comparable with that of light. The physical theory of relativity suggests, although without absolutely conclusive proof, that physical space and physical time have no separate and independent existences; they seem more likely to be abstractions or selections from something more complex, namely a blend of space and time which comprises both.

It is of course always possible to take any two things of not too dissimilar nature, and blend them into a single unity which shall comprise both. Before the advent of the theory of relativity, no one could have imagined that space and time were sufficiently similar in their natures for the result of blending them together to be of any special interest. Yet such a blend has proved to be of outstanding importance for the understanding of physics.

Any ordinary three-dimensional space may be regarded as hung around a framework of three perpendicular lines, these indicating three perpendicular directions in the space, as for instance East-West, North-South and up-down. The surveyor is accustomed to treat his perceptual space in this way, and the mathematician treats his conceptual space in the same way, except that he replaces the three perpendicular directions of the surveyor by purely mental abstractions which he usually denotes by $Ox$, $Oy$ and $Oz$. Now let us imagine the surveyor's perceptual space sliced into horizontal layers of infinite thinness—much as a skilled chef will cut a round of beef into infinitely thin slices. Any single slice, contemplated by itself, forms a mere horizontal plane which possesses extension in the East-West and North-South directions, but not in the up-down direction. If we imagine these various slices now to be laid back, one above the other, in their original positions and then welded together, we shall have reconstituted the original three-dimensional space. We may say that, in performing this last operation, we have welded verticality on to horizontality and obtained something different from either, namely a three-dimensional space.

Let us now imagine these two-dimensional slices replaced by the perceptual three-dimensional spaces of some individual $A$ at *successive instants of his experience.* Let us take all these perceptual spaces, and place them contiguous to one another in their proper order. As they are to be contiguous and not overlapping, we must imagine them all placed in a four-dimensional space before we can do this. If we now imagine them welded together, they will form a four-dimensional continuum which we may describe as the space-time unity for the individual $A$. It is a conceptual space of four dimensions, and as it is constructed out of the perceptual three-dimensional spaces of a single individual $A$, we might reasonably expect it to be private and subjective to this individual.

We can create a second space-time unity out of the perceptual spaces of some second individual $B$, which we might expect to be private and subjective to the individual $B$. The theory of relativity shows that the two space-time unities we have constructed in this way will be identical for $A$ and $B$, and so also of course for any other percipients $C$, $D$, $E$,... as well. In other words the space-time unity which we build up out of private perceptual spaces of a single individual proves to be public, and so objective. Space and time separately are private, but the blend of the two is public.

We cannot speak of right-hand and left-hand directions in ordinary space, since the right hand and left hand do not belong to space, but to an observer in space; the division of space into right-hand and left-hand is meaningless except relative to a particular observer. In the same way, we cannot speak of space and time in the space-time unity—space and time do not belong to the space-time unity, but to an observer in it. But it is the body of the observer that we want, and not his mind; a laboratory equipped with cameras and various instruments of measurement would serve our purpose just as well.

Two observers who always keep close together will have the same perceptual space, but if they are moving at different speeds, and so changing their relative positions, they will have different perceptual spaces. The theory shows that these different perceptual spaces are to be obtained by taking cross-sections of the space-time unity *in different directions*. In other words, each percipient divides up the public space-time unity into space and time in his own individual way, the mode of division depending on his speed of motion.

In the same way, to use a rather imperfect analogy, a cannon-ball may be conceived as having any number of different diameters, all pointing in different directions. It would be inaccurate to speak of any one of these as the height of the ball, to the exclusion of the others. Each one has an equal claim to be regarded as the height, and can indeed be made the height by turning the cannon-ball the right way up. But so long as the cannon-ball enters into no kind of relation with other objects, such terms as height, width and length are meaningless. In the same way, time and space are meaningless when applied to the four-dimensional continuum in the abstract. But, just as, when the ball is placed on a horizontal floor, one particular diameter immediately becomes the height of the ball, so, when we put a particular scientist or observer inside the four-dimensional continuum to measure or explore, one direction immediately becomes identifiable with his time; which particular direction it will be is determined by the precise speed at which this observer is moving.

The question now arises as to whether it is possible to divide up this unity into space and time separately in a way which shall not depend on the circumstances of individual percipients. If such a way can be found, we might identify the space and time so obtained with Newton's absolute space and time. If such a way is not found, it will not prove that no such way exists, and still less that absolute space and time do not exist; the most we could say would be that they had not so far disclosed themselves. Actually in so far as ordinary physics—i.e. physics on the man-sized scale—is concerned, no such way has so far been found, and it seems highly improbable that it ever can be. For the pattern of events is known with tolerable completeness, and must, it is found, be described in terms of the space-time unity as a whole and not in terms of its separate dimensions, which do not enter into the description at all. This might have been anticipated from the circumstance that nothing less than the unity as a whole is completely objective.

Although physics on the man-sized scale may be unable to disentangle space from time, it is still possible that atomic physics or astronomy—i.e. physics on the scale of the nebulae—may have a different story to tell. Once again, an analogy may help to explain the possibilities.

Let us imagine a race of deep-sea fish, living so far below the surface of the ocean that no ray of sunlight ever reaches them; let them be of precisely the same density as the water in which they live, so that it is just as easy for them to swim up as down; let their semi-circular canals, and any other mechanism they may have for distinguishing directions, be abolished. Such a race of beings will have no means within themselves of distinguishing directions, and if they study physical phenomena, they will find that the laws of optics, electricity, magnetism, etc. make no distinction between the different directions in space. They may then announce that nature treats all directions of space equally. Having no means of disentangling the horizontal from the vertical, they will describe different directions in a purely subjective way. Up and down will not refer to directions determined relative to the earth's centre, but relative to their own backs and bellies. They will know nothing of an objective north, south, east and west; only of subjective directions, to describe which they may use such words as fore and aft, right and left.

In this analogy, the race of fish represents physicists who study physics on the man-sized scale. The three-dimensional space in which the fishes swim corresponds to the four-dimensional space-time unity of the theory of relativity in which we exist. Man-sized nature has provided no means of dividing this into space and time separately, just as the fishes had found no means of dividing their watery space into horizontal and vertical.

Now suppose that one fish has the enterprise to swim as far as the surface of the sea. He no longer studies nature on the fish-sized scale, but on a world-sized scale. When he does this, he finds a whole range of new phenomena, and amongst them a surface, objective and fixed by nature, which at once determines up-down and horizontal directions in space in a wholly objective way.

There is still a possibility that when we leave man-sized physics for astronomical physics, we may have experiences similar to those of the enterprising fish. The hypothesis that absolute time and space do not exist brings order into man-sized physics, but seems so far to have brought something very like chaos into astronomy. Thus there is some chance that the hypothesis may not be true. Newton thought that the vast masses which occupy the remotest parts of the universe might provide a framework from which to

measure absolute rest and motion, and something of the kind seems
to be needed if the pattern of events recently revealed by nebular
astronomy is to make sense. It may be that before it can make
sense, the new astronomy must find a way of determining an
absolute time, which it will then describe as cosmical time. The
space-time unity will then be divided into space and time separately
by nature itself. Apart from this possibility, all observers stand on
the same footing, each dividing the space-time unity into his own
perceptual space and his own perceptual time.

## The Theory of Relativity

The foregoing remarks embody the main conclusions of the restricted, or physical, theory of relativity which Einstein put forward in 1905. We must always remember that this theory is a
deduction from the observed pattern of events. As the pattern can
only be expressed in mathematical terms, the theory of relativity
also can only be expressed in mathematical terms. It deals with
measures of things, and not with things themselves, and so can
never tell us anything about the nature of the things with the
measures of which it is concerned. In particular it can tell us
nothing as to the nature of space and time.

Nevertheless, as it shows the mathematical measures of space and
time to be so intimately interwoven, it seems reasonable to suppose
that space and time themselves must at least be of the same general
nature. The distinction, which many philosophers besides Kant
have drawn, between space and time as forms of perception of
external and internal experience is one which can no longer be
maintained in respect of physical space and time, although it can
for perceptual space and time.

This space-time unity of the theory of relativity figures very
prominently in the philosophical system of Alexander (1859–1938),
for he supposes that it is the primordial reality out of which all
things have evolved. He conjectures that the most primitive, as
also the simplest, kind of stuff in the world is pure space-time; out
of this various kinds of matter emerge and, gradually rising higher,
develop into life, consciousness and Deity in turn. All the con-

tinental thinkers whom we have mentioned have seen space and
time as creations of mind, but for Alexander mind is a creation of
space and time.

### Representation in Space and Time

As a final remark, science and the various materialistic philosophies have proceeded for centuries on the supposition that all
objects and all events, and indeed the whole universe, can be
arranged in space and time. Quite recently science has found that
such an arrangement is inadequate. The rays of light, waves of
sound, and the various other messengers that bring us information
as to the happenings of the outer world may quite properly be
regarded as travelling in space and time; such a representation is
self-consistent, makes sense, and gives a rational account of our
perceptions. But we shall see below that we are hardly free to
depict the events which despatch these messengers as happenings
in space and time; such an interpretation does not make sense or
lead to a rational view of the universe. We find there is something
in reality which does not permit of representation in space and
time. Thus space and time cannot contain the whole of reality,
but only the messengers from reality to our senses.

## KANT'S FORMS OF PERCEPTION AND UNDERSTANDING

Besides the two forms of perception—space and time—which we
have just discussed, Kant's fourteen mental sieves comprised twelve
*categories*, or 'forms of understanding'. There is no need to enter
into any detailed discussion of these categories, for while eleven
of the twelve may or may not be of interest to logic, they are of
no interest to science. One only makes any kind of contact with
science, and this is the category of Causality; Kant thinks that
our minds are so constituted that we see all sequences of events in
terms of the cause-effect relation.

Categories figure in a somewhat different manner in other philosophical systems. Aristotle regarded them as forms of structure,
not of the mind but of the world. For Hegel they are forms of
thought in the Absolute mind, while Alexander returns to the
Aristotelian conception of categories as forms of the world itself.

Up to the present, the conclusions of philosophy have all been reached by minds which have all been of one type—the human type—contemplating their perceptions of one and the same world. So long as there is only one type of mind contemplating one world, there can obviously be no means of deciding whether Kant's forms of perception and understanding result from the structure of the world or from the structure of the mind which perceives the world.

But we have seen (p. 43) how science has just presented us with two new worlds. The world of modern science can be divided into three fairly distinct divisions—a man-sized world in the centre flanked by the minute world of atomic physics on the one side, and the vast-scaled world of astronomy on the other. The same laws of nature prevail in all three divisions, but different aspects of them assume prominence in each, to the almost complete exclusion of all other aspects, so that we may almost regard the three divisions as three different worlds, with different sets of laws in each. But the human minds which study them are the same in each case, and so must contribute the same modes of thought to the study of each.

The two new worlds have already provided us with a testing ground for *a priori* knowledge. If this really represented some inborn quality of the mind, we should have found its assertions true in all worlds; actually most of them prove to be true only in that world which we can see and study without instrumental aid. We accordingly concluded that such knowledge was found in the human mind, not because it was born there, but as a sort of sediment left by the flow of experience of the man-sized world through our minds. Residence in the worlds of the electron or of the nebulae would have left quite a different sediment in our minds, which those of us who were rationalists would then have announced as *a priori* knowledge.

A test of a similar kind can be applied to Kant's theory of knowledge. For, as we shall see in subsequent chapters, those forms of perception and understanding which are of scientific interest—namely causality and the possibility of representation in space and time—prevail in the man-sized world, but not in the small-scale world of atomic physics which we know only through instrumental study. If they really were contributions of the human mind to nature, they would be contributed to all three worlds equally. But as they are not contributed to all three worlds, we may conclude that they are not inborn modes of human thought. Again, they would seem to be ingrained rather than inborn, not so much laws that we thrust on nature as laws that we—with our limited knowledge of the world—have allowed nature to thrust on us. We think everything can be located in space and time because the world that we perceive with our unaided senses *seems* to admit of location in space and time; the reason is not that things are so located, but that the messengers from them to our sense-organs travel through space and time (p. 139). In the same way we think we see cause and effect running through everything, because the phenomena of the man-sized world *seem* to conform to a law of causality; here again the reason is not that they do so conform, but that they obey statistical laws which produce an impression of causality on our coarse-grained organs of perception (p. 131). Our experiences of our man-sized world create in our minds habits of thought which take causality and space-time representation for granted. We cannot imagine anything else because we have never experienced anything else.

If this is so, Kant's forms of perception and understanding are not so much blinkers which restrict our knowledge of the outer world as lenses which condense our knowledge. But the knowledge they condense is knowledge only of the man-sized world, being crystallized experience of this world alone; denizens of the underworld of atoms and electrons would have had other experiences, and a Kant of this underworld, even though endowed with a mental constitution just like our own, would have produced other categories and other forms of intuition.

In any case, it is probably fair to say that all that modern philosophy retains of Kant's theories on this subject is the possibility of certain forms of thought—whether inborn or ingrained hardly matters—causing our particular type of mind to select what it does rather than something else. Our own minds contribute something to the nature they study—a view, incidentally, which dates back to Nicholas of Cusa and the fifteenth century.

Even this remnant means little, unless we concede the possibility of *a priori* knowledge about the external universe. Kant's whole theory was an *ad hoc* structure designed to remove an obvious difficulty about *a priori* knowledge, and if *a priori* knowledge passes away, the need for, and to some extent the importance of, this theory passes away with it.

At the same time *a priori* knowledge was itself, in a sense, an *ad hoc* structure designed to help metaphysics in its self-appointed task of championing the doctrines of theology. It can hardly have mattered much to Descartes or Kant whether they knew that the sum of the three angles of a triangle was 180° by having proved it in their minds, or having measured it with their instruments, or having seen it by the clear light of reason. Their primary interest was in the question of principle; they wanted to be able to claim that they were possessed of knowledge which was unchallengeable because it had not reached them through the deceitful gates of the senses. And the kind of knowledge they wanted to claim was not knowledge about triangles, but about GOD, FREEDOM and IM-MORTALITY. They wanted for instance to be able to say that, science or no science, the will was free because they saw it to be so by the clear vision of their intellects.

With the passing of this special phase of philosophy, *a priori* knowledge lost its special significance, and, apart from mathematical knowledge, few philosophers have much to say in its favour to-day; at least it is generally conceded that it is of little consequence. Yet, just when *a priori* knowledge has become discredited in philosophy, an attempt has been made to revive it in physics.

## EDDINGTON'S PHILOSOPHY OF PHYSICAL SCIENCE

We have seen how Kant thought that we ought to be able to build up a 'pure science of nature' solely by the use of the *a priori* knowledge inborn in our minds. This amounted to claiming that the world could only be of one kind—or rather could only appear in one way to us, with our minds constituted as they are. Keeping our minds as they are, the Creator could not have made the world appear different to us from what it does.

Sir Arthur Eddington also thinks that we ought to be able to build up what we may describe as a pure science of nature from *a priori* knowledge, but he thinks of this *a priori* knowledge as epistemological rather than as inborn. In other words, he thinks we should find logical inconsistencies in reaching any other conclusions about the physical world than those which the physicists have actually reached from centuries of toil in their laboratories. It should be explained that this claim applies only to the general laws of nature and not to individual objects in nature, and also that when Eddington speaks of nature, he is concerned only with nature as it appears to us, and not with an objective nature outside ourselves.

The general point of view will best be understood in terms of a specific example.

We have already seen that, if light travelled with infinite velocity, it would be a simple matter in principle to synchronize all the clocks in the universe. The method would be as simple as that of setting our watch by Big Ben, and we could call in the help of telescopes as needed. But, as light does not travel with infinite velocity, we cannot synchronize distant clocks in this way; we must allow for the time light takes to travel from one clock to another, and the theory of relativity has made it clear that the synchronization of distant clocks, if it could be achieved at all, would call for a far more elaborate technique than looking through telescopes at distant clocks.

In the years 1887–1905 a great number of experiments were performed for another purpose, any one of which might have resulted in the discovery of such a technique. But none of them did, and it is now generally accepted that the synchronization of distant clocks is an impossibility. It is not impossible in the sense in which it is impossible to fly an aeroplane at 1000 miles an hour—i.e. because our technical skill is not yet sufficiently advanced—but rather in the sense in which it is impossible to fly an aeroplane to the moon—i.e. because, *as observation has shown*, nature provides us with nothing on which we can get a hold, no resistant medium to support our aeroplane. The main result of the physical theory of relativity is usually expressed in the form that it is impossible to determine an absolute velocity in space, but might almost (not quite) equally well be expressed in the form that it is impossible to synchronize distant clocks.

As a matter of historical fact, this conclusion was reached as a generalization from a very large number of experiments. Let us, however, imagine a race of beings who know without experimenting that it is impossible ever to synchronize distant clocks—to avoid cumbersome repetition, let us agree to describe them as asynchronists. These beings would not dream of performing the whole set of experiments just mentioned, because their innate convictions would tell them the results without. If they had a Kant, he would describe this knowledge as *a priori* knowledge. If they had a Descartes, he would point out that this knowledge, being independent of all experience, could claim a higher degree of certainty than if it had been derived from a finite number of experiments, any generalization from which might be negatived by further experiments.

Now Eddington claims, in brief, that we are ourselves asynchronists, that we have knowledge in our minds as to the impossibility of synchronizing distant clocks. Like Kant he describes this knowledge as *a priori*—'knowledge we have of the physical universe prior to actual observation of it'; like Descartes he claims for it a higher degree of certainty than can be possessed by knowledge derived from experiment—'generalizations that can be reached epistemologically have a security which is denied to those that can only be reached empirically'. This *a priori*, or epistemological, knowledge is not confined to asynchronism; this is merely a somewhat trivial example. Again like Kant (p. 35), Eddington believes that 'all the laws of nature that are usually classed as fundamental can be foreseen wholly from epistemological considerations' and further that 'not only the laws of nature but also the constants of nature can be deduced from epistemological considerations, so that we can have *a priori* knowledge of them'. It follows that 'an intelligence unacquainted with our universe, but acquainted with the system of thought by which the human mind interprets to itself the content of its sensory experience, should be able to attain all the knowledge of physics that we have attained by experiment.

He would not deduce the particular events and objects of our experience, but he would deduce the generalizations we have based on them.'

Thus for Eddington knowledge of this fundamental kind results from the constitution of our minds, which are thus once again rehabilitated as law-givers to nature in the Kantian sense. We need never have built physical laboratories, except to study matters of detail; it would have been better to have delved into our own minds, where we should have found the results of all the fundamental experiments of physics, together with the values of the fundamental constants of physics. Eddington goes on to remind us that 'whatever is accounted for epistemologically is *ipso facto* subjective; it is demolished as part of the objective world'. Fundamental physics, then, tells us something about our own minds, but nothing about the outer world. To use one of Eddington's own metaphors: 'When science has progressed the furthest, the mind has but regained from nature what the mind has put into nature. We have found a strange footprint on the shores of the unknown. We have devised profound theories, one after another, to account for its origin. At last, we have succeeded in reconstructing the creature that made the footprint, and lo! it is our own.'

Eddington's claim that the fundamental laws of physics can be foreseen epistemologically would carry more conviction if he could himself establish any one of them, even the simplest, epistemologically—in other words, if he could show that there would be a logical inconsistency in believing the laws to be different from what they are. This he never does.

It seems improbable that he ever could, for surely to speak of establishing any fact of science by epistemology alone involves a contradiction in terms. Epistemology has only one tool in its armoury. This is pure logic, and before it can be applied to a scientific fact, we must define the scientific objects about which the fact is stated. We can only do this by calling upon knowledge which has been obtained empirically. In so doing we pass beyond the realm of *a priori* knowledge, and our discussion ceases to be purely epistemological.

To illustrate by a concrete case, Eddington believes it is possible to establish epistemologically that the mass of the proton must be 1847 times that of the electron. Clearly, though, he must be careful to avoid proving at the same time that the mass of the apple is 1847 times that of the orange; if his argument proves this, we shall feel suspicious of it. He can escape this danger by defining his electrons and protons in a way which makes it clear that they are not apples and oranges. Actually he neglects to do this, with the result that, in so far as his proof of the 1847 ratio is epistemological, it is equally applicable to apples and oranges.

Of course Eddington is entirely justified in assuming that *we* know what he means by electrons and protons, but what about the visiting intelligence from another universe? Will he not be in the position of the lecturee who said the lecturer had explained beautifully how the astronomers discovered the sizes and temperatures and masses of the stars, but had forgotten to explain how they found out their names? He will not know the difference between an apple and an electron until we tell him, and before we could do this, we should have to acquaint him with whole masses of laboratory knowledge, and epistemology would be left far behind. For the visitor is supposed to be acquainted only with our system of thought, and can it be seriously maintained that this includes the knowledge that the world is made up of similar fundamental particles of two and only two kinds? So far from being an innate part of our mental equipment, this is a hypothesis that did not even enter science until a few years ago (and incidentally left it again, very hurriedly, a few years later).

It is in fact necessary to build a bridge between the abstractions of epistemology and the actualities of observed phenomena; without this epistemology is left up in the air, and cannot know what it is talking about. Kant did this by introducing his *synthetic a priori* knowledge; Eddington does it by withdrawing his claim that his *a priori* knowledge· is 'knowledge that we have of the physical universe prior to actual observation of it', and writing instead that 'to the question whether it can be regarded as independent of observational experience altogether, we must, I think, answer: No'. But this admission obviously weakens his position enormously; his natural laws are no longer foreseen 'wholly from epistemological considerations', but only from a mixture—in unknown and un-knowable proportions—of these and observation, which means simply observation combined with sound reasoning. And surely this is just the ordinary procedure of all science. Eddington's laws, being no longer reached by pure epistemology, must renounce their claims to pure subjectivity, and to 'a security that is denied to those [laws] that can only be reached empirically'. They become ordinary scientific laws, obtained in the ordinary scientific way, and the only question is whether the mathematics is right or wrong.

A simple test case is provided by the finite velocity of light. We introduced Eddington's philosophy, as he himself has done, by considering the impossibility of synchronizing distant clocks. The reason why such synchronization is impossible is that light does not pass instantaneously from place to place. Those, then, who believe it is possible to prove all the fundamental laws of nature from epistemological considerations, ought to find it possible to prove in this way that the velocity of light is finite—i.e. they ought to be able to point to some logical inconsistency involved in the idea of light travelling with an infinite velocity. Eddington, however, merely dismisses the question with the statement that it is absurd to think of the speed of light as infinite—as absurd, he says, as to think of it as hexagonal or blue or totalitarian.

So long as we look at the question from the purely epistemo-logical point of view—forgetting all that experience has taught us about space, time and propagation—it is hard to find anything absurd in the idea of instantaneous propagation. Prof. A. Wolf writes that 'down to the seventeenth century [the velocity of light] had usually been regarded as infinite, and Kepler, and perhaps also Descartes, seem to have held this view. Descartes...believed that light was not a moving substance, nor a motion at all, but a tendency to motion, or a thrust exerted by the luminous body: and he supposed that this thrust, being incorporeal, required no time for its propagation.' In the same way, most people still think of the thrust of an iron bar as an example of instantaneous propagation. Newton and his contemporaries took it for granted that gravitation was propagated instantaneously; it was over a century later that the alternative possibility of a finite speed of transmission was first considered by Laplace—not because it seemed inherently probable to him, but because he wished to leave no avenue unexplored which might solve the mystery of the moon's acceleration. And when the first observational evidence (p. 63) of the finite speed of light was produced by Roemer, it was hailed as a sensational new discovery—not as confirmation of something that had been known all the time as a matter of course. Indeed, for a time it was rejected by many of Roemer's contemporaries who continued to believe in the infinite velocity of light.

All this seems to show that there can be nothing epistemologically absurd in the idea of an infinite velocity of propagation.

Even if it could be conceded that we have *a priori* knowledge that light travels with only finite velocity, it would still be a long step further to the fundamental postulates of the theory of relativity, and Eddington claims these also as *a priori* knowledge. Sixty years ago physicists were almost unanimous in imagining space to be filled with an ether through which waves travelled at the finite speed of 186,000 miles a second. This constituted a perfectly self-consistent scheme, it made sense, and explained all the phenomena as then known, so that, so far as epistemological considerations went, it was entirely eligible as a possible explanation of the phenomena; it had to be abandoned only because experiment decided against it. If these experiments had turned out otherwise than as they did—and we can easily imagine them doing so—this scheme would probably still have prevailed. This of itself gives a sufficient proof that no epistemological arguments compel the abandonment of this scheme, whence it follows that none can require the acceptance of the opposite scheme, which is that of the theory of relativity. Indeed as this latter scheme is purely a generalization from the results of a large number of experiments, there is still a possibility in principle—although not much probability—that further experiments may still be found to compel its abandonment.

### An Alternative View

There is an alternative way of regarding the matter which would seem to be more true to the facts.

Borrowing a simile from Poincaré, we have already compared

the construction of a science to the building of a house. Our stones are a collection of facts of observation. Just because nature is rational, we find that these can be made into something other than a mere shapeless pile; they show definite regularities, and so can be fitted together to form a house with definite characteristic features.

It will be possible to describe these characteristic features in simple terms which will evoke a ready response in our minds; indeed we can describe them in terms of ideas which are already in our minds and familiar to our minds. They are familiar, not because we are familiar with the general laws of physics, but because we are familiar with special and restricted instances of them; it is of such that our daily lives are made up. We may, for instance, say that the house shows no unnecessary ornamentation (Occam's razor) or no cracks (conservation laws). The ideas of ornamentation and cracks are not innate in our minds, but have been acquired from experience in very special small corners of the world.

Now the design of this house is nothing other than the pattern of events which it is the aim of physics to discover. The physicist finds—after sweat and toil in the laboratory—that this pattern of events shows features like those we have attributed to our house. There is no doubt that a great part (and perhaps all) of the fundamental facts of physics can, *when once they have been discovered empirically*, be summed up in general statements which seem very simple and intelligible to us because we are familiar with detailed instances of them. These can often (perhaps always) be expressed in the form of what E. T. Whittaker has called 'Postulates of impotence', these asserting 'the impossibility of achieving something, even though there may be an infinite number of ways of trying to achieve it'. It is, for instance, impossible to get mechanical work out of matter which is at a lower temperature than the surrounding objects, and impossible ever to measure an absolute velocity in space. These two postulates of impotence contain practically the whole contents of thermodynamics and of the physical theory of relativity respectively.

Hence, as Whittaker has remarked, 'It seems possible that, while physics must continue to progress by building on experiments, any branch of it which is in a highly developed state may be exhibited as a set of logical deductions from postulates of impotence, as has already happened to thermodynamics. We may therefore conjecturally look forward to a time in the future when a treatise on any branch of physics could, if so desired, be written in the same style as Euclid's *Elements of Geometry*, beginning with some *a priori* principles, namely postulates of impotence, and then deriving everything else from them by syllogistic reasoning.'

These principles would not of course be *a priori* in Kant's sense of 'pre-observation'; they would be very much *a posteriori*, being the highly concentrated extracts of immense masses of observations. But we can imagine a scientist pondering over their simplicity until they became endowed in his eyes with a quality of 'inevitableness', and he would begin to regard them as laws of thought. In a sense they would have become laws of thought for him.

This, we may conjecture, is what Eddington has done. And, just as the true nature of Kant's supposed categories of thought is disclosed by experiments on the atomic world, which show that causality and space-time representation no longer prevail there, so at any time a new experiment may show that Eddington's supposed *a priori* principles are mere mental sediments left over from actual experience of the world. Indeed to some extent the discovery of positrons has done this already.

## THE METHOD OF SCIENCE

Our discussion seems to bring us back to the age-old conclusion that if we wish to discover the truth about nature—the pattern of events in the universe we inhabit—the only sound method is to go out into the world and question nature directly, and this is the long-established and well-tried method of science. Questioning our own minds is of no use; just as questioning nature can tell us truths only about nature, so questioning our own minds will tell us only truths about our own minds.

The general recognition of this has brought philosophy into closer relations with science, and this approach has coincided with a change of view as to the proper aims of philosophy. The ancient philosophers pursued their studies in the hope of finding a lantern which should guide their feet along the best path in their journey through this life, the philosophers of the seventeenth and eighteenth centuries in a fixed determination to find evidence that this journey ended in a life to come. This humanistic tinge has taken a long time to disappear, but has almost done so in recent years; philosophy has become less concerned with ourselves and more concerned with the universe outside ourselves. It is now recognized that, in Bertrand Russell's words: 'Man on his own account is not the true subject-matter of philosophy. What concerns philosophy is the universe as a whole; man demands consideration solely as the instrument by means of which we acquire knowledge of the universe....We are not in a mood proper to philosophy so long as we are interested in the world only as it affects human beings; the philosophic spirit demands an interest in the world for its own sake.'

This may seem to suggest that philosophy should have not only the same methods but also the same aims and also, broadly speaking, the same field of work as science. But the distinction mentioned at the beginning of the present chapter still holds good. The tools of science are observation and experiment; the tools of philosophy are discussion and contemplation. It is still for science to try to discover the pattern of events, and for philosophy to try to interpret it when found.

# CHAPTER III

# THE TWO VOICES OF SCIENCE
# AND PHILOSOPHY

## (PLATO TO THE PRESENT)

We have seen that knowledge of the external world can come only through observation and experiment. These tell us that the world is rational—its events follow one another according to definite laws, and so form a regular pattern. The primary aim of physics is the discovery of this pattern; we have seen that it can be described only in mathematical language.

We have seen that physics cannot clothe the mathematical symbols of this description with their true physical meaning, but physics and philosophy may properly engage in joint discussion as to their possible meanings, and the most probable interpretation of the pattern of events. Yet there are many hindrances to such discussion. In the present chapter we shall try to unearth some of these and eliminate them with a view to clearing the ground for the discussions which are to follow.

## DIFFERENCES OF LANGUAGE

Foremost among these hindrances are differences of language and of terminology; when science and philosophy are not speaking entirely different languages, they often seem at least to employ different idioms.

More than three hundred years have elapsed since Francis Bacon wrote of the 'Idols' which beset men's minds when they try to discover truth. The most troublesome of these, he said, are the idols of the market-place, the place where men meet to talk with one another. For words are unsuited to the expression of accurate or scientific thought, and apparent differences of opinion often result from inadequate definition of the terms employed in the discussion.

In the intervening period science has constructed its own language, or jargon as some may prefer to call it. Unbeautiful though it may be at times, it has the great merit of exactitude; generally speaking, its terms are clearly and unambiguously defined, so that each word means the same thing to every scientist, and this thing is perfectly precise. When a physicist reads a sentence of Newton or Einstein, he may or may not understand the meaning of the sentence, but he is in no doubt as to the meaning of the words.

As science advances, new accessions to knowledge are continually being interwoven into its terminology, with the result that this continually gains in richness and precision. Here a group of new words will be necessitated by a group of new facts; there a modification in the usage of old words is called for by new knowledge of old facts. For instance the new knowledge introduced by the theory of relativity compelled us to modify our use of the words 'motion', 'velocity', 'simultaneity', 'interval of time', and so on.

There is nothing to correspond to this in philosophy, which still has no precise or agreed terminology. A great number of common words, as well as more technical terms, are used in a variety of different senses, often by the same writer. And even when philosophy uses a word in a precise and unique sense, this sense is often different from that of science.

This not only constitutes a serious hindrance to discussion between science and philosophy, but may even obscure the issue in purely philosophical problems. Indeed it is hardly too much to say that a large proportion of the puzzles and problems of the philosophy of the past owe their very existence to the imperfections of language. Many of these old problems look very different when translated into the idiom or language of science, while some vanish away in the process of translation.

There seem to be three principal causes for these differences of language and usage; it may be well to enumerate these first, and discuss them in detail, with examples, afterwards.

I. Philosophy seems to have no agreed or precise terminology because there is no agreed body of fundamental knowledge for a precise terminology to describe.

II. The language of philosophy differs from that of science largely because philosophy tends to use words in subjective, and science in objective, senses.

III. The language of philosophy further differs from that of science because philosophy tends to think in terms of facts as they are revealed by our primitive senses, while science thinks of them as they are revealed by instruments of precision.

As a preliminary to considering the first suggested cause, let us notice that science also had no agreed or precise terminology until it had something agreed and precise to describe. We have had an illustration of this on p. 25, where we saw the word 'motion' used in a very indefinite sense. Indeed three centuries ago there was a general confusion of thought between the three distinct measures that are now described as velocity, momentum and energy, and the same word 'motion' was often used to denote all three. It is the same now in those departments of science in which the fundamental facts are still under discussion; for instance, Eddington remarks that 'the terminology of the quantum theory is now in such utter confusion that it is well nigh impossible to make clear statements in it'.

A large part of philosophical terminology has always been in a corresponding state, and it may perhaps be argued that such a state is inevitable now, and will remain inevitable until philosophers can agree on their fundamental facts. Still, there can be other opinions about this. For fifty years, off and on, Leibniz was trying to devise a precise technical language and to construct a calculus for philosophy. He hoped to find that all the fundamental ideas of reasoning could be reduced to a very small number of primitive elements or 'root-notions', each of which could then be designated by a universal character or symbol like the symbols of algebra. If once this could be done, it ought to be possible to construct a calculus for the operation of these symbols. Leibniz considered that such a calculus would settle disputes between philosophers as easily as arithmetic settles disputes between accountants; if two disagreed, they would simply say, 'let us reckon it out', and sit down with their pens. But his efforts failed, and more recent attempts of the same kind have been applicable at most to small regions of the whole province of thought. The result is that philosophy still struggles to express itself in the inadequate words of common speech. It is still true, as Anatole France said, that 'un metaphysicien n'a, pour constituer le système du monde, que le cri perfectionné des singes et des chiens'.

Yet the major problems of philosophy are for the most part very difficult; many of them tax the human mind to the utmost limits of its capacity, and have baffled the most acute intellects of our race for thousands of years—indeed it is hardly too much to say that not one of them has been solved yet. In discussing these problems we have to deal with subtle and delicate shades of meaning, and to travel in fields of thought which are far removed from those of our everyday life; this would seem to demand a perfectly precise, perfectly flexible and perfectly refined instrument. Ordinary language is none of these things; it is a rough practical tool which the common man, or the unthinking savage before him, has developed from his first rough contacts with the world to express the ideas which arose out of those contacts. It would surely be an amazing coincidence if such a tool should be found suited for abstract discussions which have but little to do with the world of everyday experience. We might as well expect a surgeon to perform a delicate surgical operation with carpenters' tools—spokeshaves, chisels and hammers.

The inadequacy of popular language to express the subtleties of philosophic thought is well illustrated by the famous proposition of Descartes—*cogito ergo sum*. Descartes, believing this proposition to be true beyond all shadow of doubt, proposed basing the whole of philosophy on it. A later generation of philosophers has pointed out the inadequacy of the proposition, and their criticism is based mainly on Descartes' use of common language. For this compelled the subject of the proposition to fall into one of three clear-cut categories—*cogito, cogitas, cogitat*—or their plurals; if the thinking does not fit into one of these moulds, common language cannot express it. Anything of the nature of telepathy, for instance, is ruled out from the outset, not on the grounds that it cannot or does not occur, but simply because common language cannot cope with it; this makes thinking the prerogative of detached personalities.

But even detached personalities change with every experience; I, who have thought, am different from that other I who existed before the thought came to me. And again the tenses of language—*sum, fui, eram, ero*—are totally inadequate to express the infinite gradations of change.

Bertrand Russell says that 'grammar and ordinary language are bad guides to metaphysics. A great book might be written showing the influence of syntax on philosophy.' In illustration he mentions Descartes, who 'thought that there could not be motion unless something moves, nor thinking unless someone thought. No doubt most people would still hold this view; but in fact it springs from a notion—usually unconscious—that the categories of grammar are also the categories of reality.' We can find a more modern illustration of the same tendency in the physics of the eighteenth and nineteenth centuries. When it had become clear that light was of an undulatory nature, physicists argued that if there were undulations, there must be something to undulate—one cannot have a verb without a noun. And so the luminiferous ether became established in scientific thought as 'the nominative of the verb *to undulate*', and misled physics for over a century.

Even when philosophical writers all use a word in the same sense, their usage is often different from that of science, and this brings us to the second of our suggested causes. Until recently science has taken it for granted that there exists an objective world entirely apart from and outside our minds, and has designed its terminology for the description of such an objective world. Philosophy has never taken such a world for granted, although individual philosophers may have argued for it; on the contrary it has realized that its primary concern must be with the sensations and ideas in our minds, which suggest to us that such a world exists. Hence an obvious tendency for science to use words in an objective, and philosophy in a subjective, sense.

As examples of this difference of usage let us consider the verb *see* and the adjective *red*.

The scientist's use of the word *see* is quite definite; when he says that he sees Sirius, he means that he believes that Sirius exists outside his mind, and that rays of light which have come from

Sirius are forming an image of Sirius on his retina and thereby affecting his brain. If a drunkard says he sees purple snakes, the scientist objects that he cannot see purple snakes because there are none to see—to the scientist the essence of seeing is the passage of rays of light from the object seen to the retina of him who sees.

Many philosophers object to this. They point out that, when I say I am seeing Sirius, I am claiming to see something which may no longer exist, since it may have disappeared in the eight years which have elapsed since the light left Sirius. Bertrand Russell considers it as incorrect to say you see a star when you only see the light from it as to say that you see New Zealand when you see a New Zealander in London. He treats the case of a physiologist examining the brain of his patient in the same way; most people, he maintains, would say that what the physiologist sees is in the brain of the patient, but the philosopher must insist that actually it is in the brain of the physiologist. On this view, the drunkard can really see purple snakes in his bedroom, but the sober man can never see green snakes in the grass, because they may have gone out of existence while their light was travelling to his eyes. In brief, the philosophers consider we can only see things which are inside our heads, while the scientists, following the more ordinary use of language, consider we can only see things which are outside our heads.

The adjective *red* is used in science to describe light which possesses quite definite objective properties; these can be specified by mentioning a number of complete waves to the inch or of complete oscillations to the second—the two definitions are exactly equivalent. When light so specified falls on a normal human eye, it produces what we describe as a sensation of redness.

The mechanism by which it does this is still imperfectly understood, but appears to be somewhat as follows. The optic nerve of the human eye is a bundle of nerve fibres which terminate in the retina in the form of rods and cones. When light falls on these nerve-endings, chemical changes occur in them which send certain electric activities along the nerve fibres to the brain; these produce sensations of light or colour in the mind. The rods are stimulated by light of any colour, even though it be very faint—it is through these that we see at night or in dim light—but they produce sensations only of light and shade, and not of colour. Stimulation of the cones, on the other hand, produces definite sensations of colour. If the rods are in an unsatisfactory state, we suffer from night-blindness; if the cones, from colour-blindness.

The development of the cones is determined by certain hereditary elements which are believed to reside in a special chromosome (the $X$-chromosome), of which every man has one, and every woman two, in each cell of their bodies. In western Europe, about one man in forty started life at his conception with this hereditary element defective, and so is permanently and unalterably colour-blind; a woman is only colour-blind if she has two hereditary elements defective, so that only one woman in several hundreds is colour-blind.

Apart from man, it is believed that very few of the larger animals are endowed with colour vision; most of them see the world only as a series of contrasts of light and darkness—somewhat as we see it by moonlight. The human sensation of redness is the origin of our conception of redness as a quality, but provides only a rough test for redness; the true test is by a set of completely inanimate instruments—spectroscope, camera and photographic plate.

When a scientist says that a flower or a motor bus is red, he means that any light that they reflect is red in the scientific sense as defined above. When sunlight, which is a blend of light of many colours, falls on a red flower, the petals of the flower reflect the red constituent of the light, and this constituent alone, into my eyes so that I see the flower by red light. If I have normal vision, this produces a sensation of redness in my mind, and I say that the flower is red. If I have not normal vision, but am colour-blind to red, I shall still see the flower by light which is red in the scientific sense of the word, although my colour-blindness may result in the light appearing of a different hue, or making very little impression on my retina; I may see it as a dull, instead of a vivid, red.

But when a philosopher says an object is red, he usually means that it produces a sensation of redness in his own, or in someone else's, eye. As with the word *see* we previously discussed, the scientist applies the adjective *red* to something objective outside his head—primarily to light—while the philosopher applies it to something inside his head—primarily to a colour-sensation. Thus colour-blindness can alter colours in the philosophic sense, but not in the scientific.

## DIFFERENCES OF IDIOM

In addition to such crude and rudimentary difficulties of pure language, further difficulties originate in the different idioms employed by the philosopher and the scientist. Not only do they express their thoughts in different languages, but the thoughts themselves tend to run on different lines of rails. This seems to result, at least in part, from the third and last of our suggested causes. The philosophers still think in a way which dates back to the earliest days of their subject, to times when no instruments of measurement were available of greater precision than the five human senses; they still describe things in terms of the effects they produce on these senses, while the scientist describes them in terms of the effects they produce on his sensitive instruments of measurement. The philosopher not only speaks but thinks in subjective, and the scientist in objective, terms.

### Quantities and Qualities

One of the more obvious results of this is that the philosopher usually thinks in terms of qualities, the scientist in terms of quantities. The philosophical lecturer may be telling his audience that a lump of sugar possesses the qualities of hardness, whiteness and sweetness, while his colleague in the science room next door may be explaining coefficients of rigidity, of reflection of light and hydrogen-ion concentration—measures of the degree to which the qualities of hardness, whiteness and sweetness are possessed. While the philosophical lecturer argues on the supposition that hot and cold are incompatibles, so that no object can be hot and cold at the same time, the science lecturer discourses on temperature, which not only measures the infinite gradations of what his philosophical colleague describes as hotness or coldness, but also bridges a gulf which the latter still treats as unbridgeable.

The consequences of this can be illustrated in terms of a simple philosophical argument which had a very long innings—in different suits of clothes—lasting the 2000 years from Plato through Berkeley to Bradley. It runs somewhat as follows:

We are in a comfortable room when a man $A$ comes in from a snowstorm outside and says, 'It is warm in here'. Another man $B$ then comes in from a Turkish bath and says, 'It is cold in here'. The argument proceeds to assert that as the room cannot be warm and cold at the same time, the heat and cold cannot be real qualities of the room, but can only be ideas in the minds of $A$ and $B$. Two other men $C$ and $D$ now come in, the one from a palace and the other from an Anderson shelter, and remark respectively that the room is small and large. As the room itself cannot be large and small at the same time—so the argument runs—the largeness and smallness can only exist in the minds of $C$ and $D$; the room cannot have any quality of size in itself. By continued repetition of this argument the room can be stripped of all its qualities in turn, and as it is nothing more than the sum of its qualities (so this particular argument runs), it disappears entirely except in so far as it exists in the minds of $A$, $B$, $C$ and $D$.

The argument looks very different when it is translated into the idiom of science. When $A$ comes in he will say, 'It is warmer here than outside', while $B$ will say, 'It is colder here than in the Turkish bath'. The argument would have to proceed that a room cannot be both hotter than a snowstorm and colder than a Turkish bath—and we see at once that the attempted inference fails entirely.

Of course we cannot dispose of an argument merely by translating it into another idiom, any more than we could disprove the propositions of Euclid by translating them into French. There must obviously be more in it all than this.

The argument fails through disregarding the distinction between subjective estimates and objective measures of temperature. When it says that a room may be deemed hot and cold at the same time, it is dealing with subjective heat and cold; these, it goes on to prove, can only be ideas in the minds of $A$ and $B$. But here it suddenly swings over and erroneously identifies them with objective temperatures. The subjective room may be the sum of its subjective qualities, and the objective room the sum of its objective qualities, but abolishing all the subjective qualities of the room cannot abolish the objective room. Before his argument can stand, the philosopher must show that there is no difference between the subjective and objective temperatures of a room, and every time he tries to do this the thermometer on the mantelpiece will prove him wrong.

The psychologist may put in a word here, since he can tell us that our senses are not very good at estimating absolute heat and cold; we do not judge that an object is hot or cold, so much as that it is *hotter than* or *colder than* something else, the comparison usually being with the warmth of our own bodies or with our last experience with heat or cold. Thus common language speaks of marble as cold and of woollen blankets at the same temperature as warm, because touching marble makes our hand *colder than* it has been and wrapping it in a blanket makes it *hotter than* it has been; the ultimate reason for this is that marbles are good, and woollen materials are bad, conductors of heat. The psychologist knows from his laboratory experiments that considerations such as this are important, while the philosopher of the old-fashioned type apparently did not. Science knows from its observations that its own idiom is the correct one to employ.

Since the time of Aristotle, philosophers have been inclined to regard substance as something that is wrapped up in a number of qualities, much as a package may be wrapped up in a number of layers of paper, and have speculated as to what, if anything, will be found when all the wrappers are removed.

Galileo, Descartes, Locke and others imagined that qualities could be divided into an outer layer of what Locke described as *secondary* qualities—those perceived by the senses, such as redness and coldness—and an inner layer of *primary* qualities which a substance or an object possesses in its own right and by virtue of its mere existence, independently of whether it is perceived or not—such as solidity and extension in space; these, in Locke's words, 'are utterly inseparable from the body in what state soever it be'.

Looked at from the objective viewpoint of science, such a distinction appears highly artificial. Redness indicates a capacity for reflecting red light, solidity and extension in space a capacity for 'reflecting' any other body which tries to trespass upon the space of the body in question. It is not clear why one of these capacities should be classified as primary and the other as secondary, one as fundamental and the other as superficial.

The philosopher may protest that to him redness has nothing to do with the reflection of light, but means simply a capacity for producing a mental sensation of redness. This will not do, since it makes the distinction between primary and secondary qualities purely subjective. Redness must now be classified as a secondary quality for a normal man, but as a primary quality for a blind man, who cannot see at all, as also for a dog, who has no colour vision. Locke and his fellow philosophers may argue that redness is a secondary quality, but a canine philosopher would argue, with precisely equal validity, that it was a primary quality.

The problem is sometimes approached by imagining an object to be stripped, one by one, of all the qualities which we can imagine being stripped from it. The qualities which we can imagine removed are of course secondary, the unremovable residuum primary. The philosophic lump of sugar, for instance, is pictured as wrapped up in its qualities of whiteness, sweetness, hardness and so forth. If we strip these away, one after the other, what unstrippable residue is finally left? Or is nothing left? Is it true, as was assumed in the argument just quoted, that an object is nothing but the sum of its qualities?

Science finds that the qualities of a substance or object depend in part on the intrinsic nature of its constituent parts and in part on the way in which these constituent parts are arranged in space, its physical qualities depending on the mode of arrangement of its molecules, and its chemical qualities on the mode of arrangement of the atoms of which its molecules are formed. This being so, it is meaningless to speak of 'stripping' anything of its qualities. The most we can do is to rearrange its constituent units, and in so doing replace one quality by another—for instance, the hardness of ice by the liquidity of water or the compressibility of steam, the brilliance of diamond by the heavy dullness of graphite or by the deep blackness of lampblack. To the scientist all qualities are primary in the sense that they 'are utterly inseparable from the body in what state soever it be'; a red tulip is not made less red by being looked at in a blue light.

Again it is not to the point for the philosopher to protest that the scientist insists on looking at things objectively, while he, the philosopher, is accustomed to keep his thoughts on the subjective plane. If he insists that he can easily *imagine* things stripped of their qualities, the reply is that philosophy, just as much as physics, is out to obtain knowledge about the real world, and not about an imaginable but wholly unreal world in which qualities can be stripped away and nothing left in their place; it is only in Wonderland that a cat can be stripped of everything but a grin.

*Half-tones*

A second difference of idiom, closely connected with that we have just discussed, arises out of the philosophical practice of depicting the world entirely in black and white, and so ignoring all the half-tones, gradualness and vagueness which figure so prominently in our experience of the actual world. The obvious example of this is provided by the law of the excluded middle, which has dominated formal logic, with devastating results, from the time of Aristotle on. The law asserts that everything must be either $A$ or not-$A$, whatever $A$ may be. The scientist, on the other hand, knowing that everything will generally possess some $A$-ness, and some not-$A$-ness, is very little concerned as to whether an object is classed as $A$ or not-$A$; what he wants to know is how much $A$-ness it possesses.

For example, the law asserts that every quantity must be either finite or not-finite. If this is so, the half of a finite quantity must always be finite; it cannot be not-finite, or the sum of two not-finite quantities would be finite, which is absurd. Thus in the series of quantities

$$1, \tfrac{1}{2}, \tfrac{1}{4}, \tfrac{1}{8}, \tfrac{1}{16}, \tfrac{1}{32}, \ldots,$$

in which each is half of the preceding, every member of the series must be finite no matter how far the series extends. If it continues indefinitely, we have an infinite sequence of quantities each of which is finite. The sum of all the members of the series is now the sum of an infinite number of finite quantities, and so must, according to the law, be infinite. Yet very simple arithmetic will show that the sum is actually finite, being 2.

This is the fallacy underlying Zeno's well-known paradox of the hare and the tortoise. For simplicity, let us suppose that the hare goes only twice as fast as the tortoise. Let it give the tortoise a start of a minute, during which the tortoise travels from the starting-point $A$ to a point $B$. The hare now starts, and takes half a minute to reach $B$. During this time the tortoise travels a distance $BC$, which is of course half of the distance $AB$. The hare accordingly takes a quarter of a minute to traverse the stretch $BC$. And so it goes on, the total time of the race, in minutes, being

$$1 + \tfrac{1}{2} + \tfrac{1}{4} + \tfrac{1}{8} + \dots \ ad\ inf.$$

Obviously the series can never end, and as, according to the law, it consists of an infinite number of finite terms, the total time of the race must be infinite—the hare can never catch the tortoise. As before, the fallacy lies in the supposition that quantities can be sharply divided into finite and not-finite—in other words, in the law of the excluded middle.

To turn to a more serious example, the same fallacy lies at the root of the so-called ontological proof of the existence of God. In the form in which St Anselm originated it, this assumed that a being, like an object, must either possess or not possess every conceivable quality. Thus a Perfect Being must either possess or not possess the quality of existence; He must, in fact, possess it, since the non-possession of this quality would be an obvious imperfection. Hence, runs the argument, a Perfect Being must really exist. The detailed argument, in the form given by the usually clear-minded Descartes, ran as follows: 'To say that an attribute is contained in the nature or in the concept of a thing is the same as to say that this attribute is true of this thing, and that it may be affirmed to be in it. But necessary existence is contained in the nature or in the concept of God. Hence it may with truth be said that necessary existence is in God, or that God exists.'

We can almost see the rabbit being put into the hat, and it seems strange that such a transparent piece of logical legerdemain could impose not only on the confused logicians of the Middle Ages, but also on later thinkers of the calibre of Descartes and Leibniz, until Kant finally pointed out its logical inadequacy: 'that unfortunate ontological argument, which neither satisfies the healthy common-sense of humanity nor sustains the scientific examination of the philosopher'.

The explanation seems to be that only two different degrees of existence were recognized—existence and non-existence. The argument proves that if we set out to think of a Being endowed with every perfection, we must think of Him as really existing— nothing more. It can never assign a higher degree of existence to such a Being than existence in our thoughts—*ex nihilo nihil fit*.

As soon as the argument is translated into the scientific idiom, we are no longer concerned with mere qualities, but with degrees of qualities, and if the being is to be identified with the Supreme Being, the degree of each can only be infinite. But, as Leibniz pointed out, there are pairs of qualities which become incompatible when taken in infinite amounts, as infinite justice and infinite mercy. Thus, so far as this argument goes, we have no right to imagine such a Supreme Being, even in our thoughts.

The law of the excluded middle entails other disconcerting consequences of a more practical kind. From it we learn that at every moment of his life a man must be either young or not-young, so that the transition from young to not-young must occur at a single moment of his life. Thus youth passes away in the twinkling of an eye, and it is the same with the beauty of a woman and the health of an invalid. We reach strange conclusions by following the strait path of formal logic.

In practical affairs all life is a compromise, and most things reside in precisely that middle region which the law attempts to abolish. This does not in the least interfere with the popularity of the law for dialectical purposes: 'Gentlemen, it is surely obvious that there is either a shortage of feeding stuffs for pigs, or there is no shortage.'

## DIFFERENCES OF METHOD

A natural transition of thought brings us to a third difference of idiom, or perhaps rather of method, which has somewhat more serious consequences than any so far considered. The *métier* of the philosopher is to synthesize and explain facts already known; that of the scientist is in large part to discover new facts. When the philosopher finds himself called upon to explain a very complex and very unintelligible world, he is tempted to reduce every problem to its crudest and barest skeleton by discarding everything which does not seem to him to be essential. The scientist, on the other hand, ever looking for something new, naturally preserves all complications; indeed he welcomes them, since they may show him the way to new fields of knowledge. The point of interest to us at the moment is that the philosopher is in danger of over-simplifying his problem, and leaving out essentials through not seeing that they are essential.

### Over-Simplification

To take a simple illustration, the philosopher may set out to inquire why a flower looks red in the philosophic sense—wherein does its philosophic redness reside? Like so many of the fundamental problems of philosophy, this dates back to Plato; in the *Theaetetus*, Socrates reaches the conclusion that colour resides neither in our eye nor in the perceived external object. The modern philosopher usually follows the lead of Plato to the extent of eliminating all factors from the discussion except the flower and the mind which perceives it, for surely these and these alone (so he will say) are the essentials of the problem. He can now argue that to one mind the flower may appear crimson and to another scarlet; hence the colour cannot reside in the flower; hence it must reside in the perceiving mind; and so on, as on p. 90.

The scientist knows how many other factors are involved. In particular, the light by which the flower is illuminated must be important, since if there were no illumination, the flower could not look red at all—it would look black. Actually it cannot look red unless there is some red light to be reflected, so that there must be a red constituent in the light by which the flower is illuminated. And even if there is red light to be reflected, a man will not see this unless his retina is sensitive to red light, so that he must not be colour-blind to red. Thus we see that for a flower to look red, three conditions must be fulfilled:

(a) The illumination of the flower must contain some red light.

(b) The surface of the flower must have the power of reflecting red light.

(c) The man who looks at the flower must not be colour-blind to red.

The question as to where the philosophic redness of the flower resides no longer seems to be expressed in the best form, but if an answer must be given, it should clearly be that the redness resides in

(a) The sun or some other illuminant which emits red light.

(b) The surface of the flower which reflects red light.

(c) The retina of the percipient which perceives red light.

This brief discussion will have shown that the perception of redness is far more complicated than the simple treatment of the philosopher usually assumes, and even so it is still far from covering the whole ground.

If, instead of asking why a flower looks red, we ask why the setting sun looks red, the answer just given fails entirely. The new answer is that the earth's atmosphere abstracts certain constituents from the sunlight as this passes through it; that it abstracts more blue light than red, making the blueness of the sky therewith; that this abstraction increases the proportion of red in the remaining light, so that the sun *always* looks redder than it really is. But at sunrise or sunset the sunlight makes a longer journey than usual through the atmosphere, so that more than the average amount of blue light is abstracted, and the sun looks even redder than usual; comparing it with its ordinary appearance, we say the sun looks red.

To put it in another way, a long process of evolutionary development has given our race eyes which are sensitive only to those wave-lengths of radiation with which the sun mainly lights the earth, and are most sensitive to those which arrive in greatest profusion. At sunset the normal balance of these colours is disturbed in the way just explained, and sunlight looks red.

If again we ask why the most distant objects in space look red, as they all do, we come into contact with one of the big outstanding problems of present-day astronomy. The objects in question are the great extragalactic nebulae, and they do not reflect light as a flower does, but emit light of their own. The more distant a nebula, the redder its light. It may be that the light would appear yellow, green or blue to an inhabitant of the nebula, and that it looks red to us only because we are receding from the nebula (or the nebula from us, which comes to the same thing) at a speed comparable with that of light. This would result in the light-waves entering our eyes at less frequent intervals, and this in turn would cause the light to appear redder to us than to an inhabitant of the nebula. But there are other possibilities, too technical for discussion here.

Other colour-problems, with entirely different answers, are provided by the redness of a fire, the blueness of the electric arc, the blueness of the sky (partially explained above), the blueness of moonlight and of shadows on the snow, and by the varied colours of the rainbow, of the butterfly's wing and of the patch of dirty oil on the road. But whether we discuss the colours of the rose or of the butterfly, of the nebula or of the rainbow, the philosophers must concede that there are more things in heaven and earth than are dreamed of in their philosophy; the world is not as simple as they try to make it.

## Atomistic Modes of Thought

Another difference of method is that the philosopher is much more 'atomistic' in his thought than the scientist. He is inclined to see the world as a collection of separate objects, nature as a collection of detached events, time as a collection of moments each of finite duration, and space as a collection of regions each of finite extent. The scientist, on the other hand, thinks mainly in terms of continuity. He sees nature as a theatre of continuous change rather than as a succession of jerks, as a cinematograph show rather than as a series of magic-lantern slides. While the philosopher thinks of time as a succession of finite moments, the scientist represents it as an ever-flowing stream; if he divides it into moments, each is infinitesimal in size, so that the time interval between two successive moments is nil. It is the same with space; the philosopher divides this up into small finite regions, but the scientist into infinitesimals or points, the distance between two again being nil. In brief, the philosopher tends to think in terms of what the mathematician calls *finite differences*, whereas the scientist thinks in terms of *infinitesimals*.

Possibly this last remark not only summarizes the difference, but also explains its origin, which seems, at least in part, to be historical. The modes of philosophical thought had become crystallized before Leibniz invented the differential calculus or Newton the theory of fluxions. As science progressed to ever new types of problems, the scientist had perforce to acquaint himself with the newer and more accurate modes of thought or fail in his attack, whereas the philosopher, still concerned with the same old problems, experienced no such need. There are of course exceptions. Leibniz, as was to be expected from the inventor of the differential calculus, always insisted strongly on the continuity of all change in nature, as has also Bergson in more recent times.

The question is one of more than mere form. There is a common belief that discontinuous change inevitably passes over into continuous change if the intervals of the discontinuities are made vanishingly small. In some respects this is true, but in others it is not. No matter how small we make its steps, a staircase will never become the same thing as an inclined plane. A sufficiently small particle can always stand at rest on the staircase, but will roll down the inclined plane; more paint is needed to paint the staircase than to paint the inclined plane—41 per cent more if the angle is 45°, regardless of how large or small the steps may be. Again, a saw is not turned into a knife by making its teeth infinitely small; the two will cut their way through matter by quite different processes.

An example of this atomic mode of thought and its consequences is provided by another of the famous paradoxes of Zeno. Imagine that a moving arrow has some position $P$ in space at some moment $A$, and some other position $Q$ at the next moment $B$. If we regard time as a succession of separate moments $A, B, C, \ldots$, there must be some instant of time at which the moment $A$ gives place to the moment $B$, and this instant is common to the moments $A$ and $B$. Because it belongs to $A$, the arrow must be at $P$ when it occurs, and, because it belongs to $B$, the arrow must be at $Q$. But it is impossible for the arrow to be at two different places $P$ and $Q$ at the same instant of time, so that $P$ and $Q$ must be the same, which means that in the time-interval from $A$ to $B$ the arrow cannot have moved at all. In this way, Zeno claimed to prove, although perhaps with his tongue in his cheek, that all motion was impossible and all change an illusion. Reality must then be changeless, the doctrine which Parmenides had set up in opposition to the πάντα ῥεῖ, καὶ οὐδὲν μένει of Heraclitus.

When this argument of Zeno is translated into the scientific idiom nothing of it is left. As the interval between two successive moments is now nil, there is no significance in the motion of the arrow in this interval also being nil. To come to grips with the problem, we must consider the motion of the arrow throughout an infinite number of moments, since nothing less than this will give us a finite interval of time. The distance through which the arrow moves in an infinite number of these infinitesimal moments is of course

$$\text{infinity} \times \text{zero},$$

which, as every schoolboy knows, may be zero or finite or infinite. Thus the possibility of motion is re-established, and the universe again becomes free to change.

When the philosophers of a later age came to study problems of motion and change, a large part of their arguments was vitiated by their habit of still dividing time into detached moments and change into detached events; it was as though they could see nothing in the Great North Road except a succession of milestones. Neither Kant nor Berkeley seems ever to have grasped the general principle of infinitesimals, the latter protesting that it had been 'contrived on purpose to humour the laziness of the mind, which had rather acquiesce in an indolent scepticism than be at the pains to go through with a severe examination of those principles it hath ever embraced for true'. Maintaining, as he ever did, that existence consisted in being perceived, he indignantly refused to admit that infinitesimals could exist which were too small to be perceived, or that mathematicians could stand to gain by imagining them to exist when they did not. He was especially severe on those who 'assert that there are infinitesimals of infinitesimals of infinitesimals, without ever coming to an end. So that according to them an inch doth not barely contain an infinite number of parts, but an infinity of an infinity of an infinity *ad infinitum* of parts'.... 'Whatever mathematicians may think of *fluxions* or the *differential calculus* or the like, a little reflection will show them that, in working by those methods, they do not conceive or imagine lines or surfaces less than what are perceivable to sense. They may indeed call those little and almost insensible quantities *infinitesmals* or *infinitesimals of infinitesimals*, if they please: but at bottom this is all, they being in truth finite, nor does the solution of problems require the supposing any other.'

### Causality

The results were particularly disastrous in discussions of the problems of causality. Many philosophers imagined that the happenings in nature could be broken up into isolated events, and that these could be grouped in pairs, in such a way that the events of each pair were related through the cause-effect relation.

On this fallacious basis, Kant argues that 'the greater part of operating causes in nature are simultaneous with their effects', for the reason that 'if the cause has but a moment before ceased to be, the effect could not have arisen'. He instances a warm room which, he says, is warm because a fire *is burning* in it, although, as every housemaid knows, the reason is that a fire *has been burning* in it.

Kant sees that, if cause and effect really are simultaneous, it becomes difficult to say which of a pair of related events is cause and which is effect, but claims to be able to distinguish between the two 'through the relation of time of the dynamical connection of both'. To take his own illustration, a leaden ball lying on a cushion is invariably accompanied by a hollow in the cushion. 'If I lay the ball upon the cushion, then the hollow follows upon the previously smooth surface; but if the cushion has for some reason or other a hollow, there does not follow thereupon a leaden ball.'

Hume subsequently propounded a different view of causality, holding that all effects are contiguous in space with their causes, and also successive in time. But contiguity and succession are not enough of themselves to proclaim two objects or events to be cause and effect; there must also be *constant conjunction*. In other words we must have noticed the contiguity and succession repeated in a great number of instances. 'We remember to have seen that species of object we call flame, and to have felt that species of sensation we call heat. We likewise call to mind their constant conjunction in all past instances. Without any further ceremony, we call the one *cause*, and the other *effect*, and infer the existence of the one from that of the other.' This also is very unconvincing scientifically, partly because heat is frequently experienced without flame, and flame without sensible heat; partly because there is no means of deciding which is cause and which effect. In actual fact heat often produces flame and flame usually produces heat, but when we come upon a house on fire, it is not easy to say whether the ultimate origin of the conflagration was heat or flame or something different from both.

Obviously, too, the constant conjunction of two events does not entitle us to ascribe the cause-effect relation to them at all. I may remember having repeatedly seen the Scotch express pass through my station when the hands of my watch pointed to 12 o'clock, but this does not show that either event is the cause of the other. We may have repeatedly seen the full moon when the sky is clear, and never when the sky is clouded over, but must not conclude that the full moon makes the sky clear (although there is a popular superstition to this effect), or that a clear sky makes the moon full.

Typical of a more modern and more scientific attitude to causality is the definition recently proposed by Bertrand Russell: 'Given an event $E_1$, there is an event $E_2$ and a time-interval $T$, such that whenever $E_1$ occurs, $E_2$ follows after an interval $T$.' Yet scientific study shows that even this is not true, to the exactness which philosophy ought to aspire to, except in the one special case in which $E_1$ is the state of the whole world at one instant of time, and $E_2$ is the state after a time-interval $T$.

The scientific fact is that it is not permissible to treat the causal relation in any of these ways. All are based on unwarranted simplifications of the complexities of the actual world; they are abstractions which can at best provide approximations to the truth.

There is no scientific justification for dividing the happenings of the world into detached events, and still less for supposing that they are strung in pairs, like a row of dominoes, each being the cause of the event which follows and at the same time the effect of that which precedes. The changes in the world are too continuous in their nature, and also too closely interwoven, for any such procedure to be valid. We shall see this more clearly when we discuss the scientific view of causality in the next two chapters, but it may be of service to illustrate it by a simple example here and now.

Suppose I shoot a bird and it falls to the ground. The falling to the ground may obviously be regarded as an effect, but where are we to look for the cause? In spite of Kant's argument to the contrary which has just been mentioned, most men would say that it was my having previously pulled the trigger of my gun. Yet this is an obvious over-simplification of the situation; to my pulling the trigger must be added my having previously loaded the gun with a cartridge in which someone had previously put powder and shot in the right places and in the right amounts, that I further pointed the gun in the right direction, and pulled the trigger at the right moment, having previously made a correct allowance for the speed and direction of flight of the bird, for the strength and direction of the wind, and for the effects of air resistance and gravitation. That the shot found its mark when I aimed in this particular direction was perhaps because a depression which had been centred over Iceland three days ago had moved eastward and caused strong south-west winds; this was because there had been a hurricane in the West Indies a week before, and so on *ad infinitum*. Any effect is seen to be connected to previous events by an endless succession of strings of events all of which meet in the effect.

We see how excessively naïve it is to suppose that all the events in the world can be arranged in pairs with the cause-effect relation obtaining in each. This would imply that each effect has only one cause and each cause only one effect. If we suppose that the happenings of nature are governed by a causal law, we must suppose that the cause of any effect is the whole previous state of the world, so that every effect has an infinite number of causes. Some of these may of course exert an influence so slight as to be negligible. For instance, my success in hitting my bird will not depend to any appreciable extent on whether Sirius is in the ascendant, or whether I have just broken a mirror or spilt the salt, although it may depend on how late I sat up the night before.

Yet in considering any event, it is not necessary for all previous events in the history of the world to be considered as separate causes. The effects of the earlier of them are already taken into account in the later, and they need not be allowed for twice over. It is enough to consider a cross-section at one particular instant of time. The state of the world at this instant—any instant I choose— will provide the adequate cause of the effect under consideration. If for example I select the instant at which I pulled the trigger to shoot the bird, then the state of the world at this instant already comprised a cartridge in my gun and a strong south-westerly wind; there is no need for us to bother about who loaded the gun or what caused the wind.

The cross-section we select need not extend over the whole of space; the more distant regions may be left out of consideration altogether. For no influence can travel faster than light, and some parts of the universe will always be so distant that light which left them at the instant of the cross-section would not have reached us yet; happenings at such places obviously cannot affect the present course of events here.

Two particular cases of cross-section are of special interest. First, the cross-section may be taken at the beginning of time, or, if we prefer so to call it, at the creation of the world; we then see how everything that occurs now is a direct consequence of the way in which the atoms of the world were arranged at their creation. Second, we may bring our cross-section forward through time until its instant differs only infinitesimally from the present. All those parts of the universe which are not in our immediate vicinity may now be disregarded, and we find that the state of things here and now depends only on the state of things which prevailed in our immediate vicinity an infinitesimal moment ago. This brings us back to the very restricted view of causality adopted by Kant, but science sees no reason for confining itself to this view. Neither does the common man, who will continue to insist that his dog died to-day because it had eaten poison yesterday.

# CHAPTER IV

## THE PASSING OF THE MECHANICAL AGE

### (NEWTON TO EINSTEIN)

#### PRE-NEWTONIAN MECHANICS

The earliest attempts to discover the pattern of events were limited, naturally enough, to the visible movements of objects either on what we have called the man-sized scale or on the far grander scale of astronomy—these were the only movements which could be studied without instrumental aid.

The movements of the astronomical bodies were treated only in their geometrical aspect. The 'fixed stars' hardly came under discussion at all, since they appeared to have no motion beyond their diurnal rotation round the pole. This was of course a consequence of their great distance from the earth, but it was explained by supposing them to be immovably attached to a sphere which rotated round the earth as centre.

There remained the sun, moon, and planets. A whole succession of astronomers—from Aristarchus through Ptolemy to Copernicus and Kepler—had investigated the paths in which these bodies moved, but had shown very little concern as to why they moved in these particular paths rather than in others. Aristotle's pronouncement that a circular motion was natural to all bodies, because the circle was the perfect geometrical figure, seems to have stifled curiosity fairly thoroughly for nearly two thousand years; it was uncritically accepted by Copernicus, and even at one time by Galileo.

It was different with terrestrial bodies; there had been many attempts to explain their movements in what we should now describe as dynamical terms. The earliest Greek thinkers had imagined that the motion of every object was controlled by a tendency, inherent in the object, to find its 'natural place' in the world. A stone sank in water because the natural place for stones was the bottom of a stream; flames ascended in air because their natural place was up in the sky, and so on. Aristotle explained this by the supposition that bodies possessed varying degrees of heaviness and lightness, and that the natural arrangement of the world was one in order of heaviness, the heavier bodies taking their places below and the lighter above—like layers of oil and water. This view prevailed until it was challenged by Giordano Bruno (1548–1600), who pointed out that as heaviness and lightness were merely relative terms, substances could have no natural places in the universe.

It was of course obvious that many objects were not in their natural places, and some explanation had to be found for this. Aristotle had thought that a body could only be kept away from its natural place through continued contact with some other body, such as the hand which held it or the table on which it lay; it could only be moved by the pressure of some other body, and this contact had to persist throughout the motion. When a stone was thrown upwards, the air surrounding it was also set into motion and pressed on the projectile through its flight, thus keeping it from returning to its natural place which was on the ground.

A different view was held by Hipparchus (c. 140 B.C.), who thought that a body was set into motion by receiving an 'impulse' from some other body; this stayed in the moving body for a time, but then gradually weakened and finally disappeared, with the result that the moving body first slowed down and finally came to rest.

It was natural that such beliefs should be held, for they seemed to be confirmed by the actual behaviour of moving bodies at the surface of the earth. Here every moving body obviously did slow down and finally come to rest; had it done anything else it would have formed a perpetual-motion machine, and this was generally agreed to be an impossibility. Indeed Aristotle had branded it as an absurdity, using it thus in an argument which ended in a supposed *reductio ad absurdum*. But the true reason for the slowing down was not that conjectured by Hipparchus; it was the action of air resistance, friction and other 'dissipative' forces.

A first glimmering of the truth seems to have been seen by Plutarch, who wrote (c. A.D. 100): 'Everything is carried along by the motion natural to it, if it is not deflected by something else.' Apart from this, none of the ancients seems ever to have conjectured that a body set in motion in empty space, or in any region in which dissipative forces did not operate, would not slow down at all, but would really act as a perpetual-motion machine and continue to move, either for ever or until something extraneous brought it to rest.

The idea that such motion could occur is, however, definitely found in the writings of Nicholas of Cusa (1401–1464). He believed that the earth continually moved through space without our being conscious of its motion—just as a boat may drift down a river without its occupants knowing they are moving until they notice the banks sliding past them—and also accepted the Pythagorean doctrine that the earth turns steadily on its axis once every twenty-four hours. He further remarked that a smooth ball which has been set moving on a smooth floor will continue to move until something checks its motion. Here his facts were right, but his explanation wrong; he supposed that the motion continued because every particle of the ball tended to retain its natural circular motion round the centre of the ball, remarking that a ball which was not perfectly round would not persist in its motion.

Then came Galileo, who saw that the primary effect of outside influences acting on a body is to produce a change in the *motion* of the body, changes of position being only secondary effects. Thus a body which is not acted on by any outside influence at all can experience no change of motion, and so must move on for ever at the same uniform speed, as Nicholas of Cusa had said.

Descartes was probably the first to enunciate this principle clearly and unambiguously, writing: 'A body when it is at rest has the power of remaining at rest, and of resisting everything that could make it change. Similarly when it is in motion it has the power of continuing in motion with the same velocity and in the same direction.'

Descartes was also the first, at least since the era of Greek speculation, to attempt to bring all the phenomena of physics within the scope of a single system of laws. His system was not dynamical but kinematical; he tried to explain phenomena in terms of motion and not of forces: 'I do not accept any other principles in physics than there are in geometry and abstract mathematics, because all the phenomena of nature may be explained by their means.' But the system was mostly erroneous.

By contrast, the system which Newton published in the year 1687 under the title *Philosophiae Naturalis Principia Mathematica* was purely dynamical in its nature. If it was not perfectly true to nature, it was at least so true that two hundred years were to elapse before its imperfections began to show themselves.

## THE NEWTONIAN MECHANICS

Newton regarded the material world as a collection of particles or pieces of matter, each of which could be either at rest in space or moving through space. If a particle was at rest it stayed at rest, and if it was in motion it continued in motion—at the same speed and in the same direction—unless 'forces' intervened to change this state of rest or motion (First Law). Thus perpetual motion became the normal state of things for a moving body unless something checked it.

Forces were explained only by their effects, which were to change motion; a force was measured by the change it produced in the velocity of the body on which it acted, multiplied by the mass of the body (Second Law). Here the word 'velocity' must be understood to specify not only the speed but also the direction of the motion. Thus a change of velocity must be supposed to occur when a body changes the direction of its motion, even though it continues to move at the same speed—as with the moon's motion round the earth; the force which causes this change of velocity is of course the gravitational pull of the earth.

Newton added that when any body *A* exerts a force on a second body *B*, then *B* must exert on *A* a force which is equal in amount but opposite in direction (Third Law).

For two reasons Newton's system of mechanics was incomparably better than anything that had preceded it. In the first place, it was based on the experimental results of Galileo and others, whereas previous systems had been based on conjecture and speculation. And, in the second place, it was free from any special concern with the local conditions prevailing at the earth's surface, and so was able to provide a sound basis for the vast superstructure of dynamical astronomy which was subsequently to be reared upon it—it was a dynamics for the sky as well as for the earth. Yet it represented only one step, although an important one, towards final truth. For underlying it was the assumption that bodies moved against a background of absolute time and space; two hundred and thirty years later the theory of relativity was to disclose that nature provides no such background. And after another ten years the theory of quanta was to show that Newton's laws are valid only for the larger scale phenomena of nature; beyond these lies a whole world of atomic and sub-atomic processes which do not obey Newton's laws at all.

### Mechanistic Determinism

This system of mechanics threw into perfectly sharp focus the problem of determinism upon which we touched at the end of the preceding chapter. According to Newton's laws, any particle $A$ in the world will be subject to forces from any or all of the other particles $B, C, D, \ldots$ in the world. These forces may come from contiguous particles—as when two billiard-balls collide—or from distant particles through gravitational attraction—as when the sun and moon raise tides on the ocean. In either case the amount of force exerted at any instant depends only on the positions which the various particles of the world occupy in space at that instant.

It follows that the changes of the world at any instant depend only on the state of the world at that instant, the state being defined by the positions and velocities of the particles; changes in position are determined by the velocities, and changes in velocities by the forces, which in turn are determined by the positions.

If, then, we know the state of the world at any one instant, we can in principle calculate to the last detail the manner and rate at which this state will change. Knowing this, we can calculate the state at the succeeding instant, and then, using this as a stepping-stone, the state at the instant next after, and so on indefinitely. Thus, as Laplace pointed out in his *Essay on Probability* (1812), the present state of the world may be regarded as the effect of its antecedent state, and also as the cause of the state that is to follow.

He went on to say that if the state of the world at its creation were specified in its minutest details to an infinitely capable and infinitely industrious mathematician, such a being would be able to deduce the whole of its subsequent history. 'Nothing would be uncertain for him; the future as well as the past would be present to his eyes.'

Even though no such mathematician exists, the whole future history of the world must have been implicit in its configuration at its creation; its so-called evolution is a mere unrolling of what is already there, and we have as little power to affect the pattern of things to come as a man who weaves a carpet on a loom which is already set, or indeed as a man who unrolls an already woven carpet for our inspection.

When once this evolutionary point of view has been gained, it becomes a mere question of words whether we speak of 'causation' with Kant or of 'constant conjunction' with Hume. If the pattern of the world is such that after $A$ always comes $B$, who shall care whether we say that $A$ is the unvarying cause of $B$, or that $B$ is the unvarying concomitant of $A$? The true and indisputable cause of everything was the arrangement of the particles of the world at the beginning of time, so that it is true to say, in the language of orthodox theology, that all things were fore-ordained by God at the creation of the world. But it is equally true to say, in the language of science, that the cause of everything is to be found in the arrangement of the particles of the world at any past instant in its history that we may choose; every past instant may equally well be treated, for our present purpose, as the moment of the world's creation. And what is essential is the arrangement of the particles, and not the God Who arranged them.

### General Principles

Although it would need Laplace's infinitely industrious and infinitely talented mathematician to trace out the future of every particle in the universe, yet quite ordinary mathematicians have been able to obtain a good deal of simple but important knowledge about the motions of particles in general.

The *kinetic energy* of a moving particle is defined to be half of the mass of the particle multiplied by the square of its speed of motion

($\frac{1}{2}mv^2$), this being the amount of work that must be done to set the particle into motion at a speed $v$. When two or more particles affect one another's motion by contact or impact, it is easily shown that any increase in the kinetic energy of one is exactly offset by an equal decrease in the kinetic energy of the others, so that the total kinetic energy of all the particles remains constant throughout the interaction.

Again the *momentum* of a moving particle is defined to be the mass of the particle multiplied by its speed of motion ($mv$). When two particles act on one another, the momentum of both is changed. If the motion is confined to one direction in space, it is easily shown that any gain of momentum by one particle is offset by an equal loss to the second, so that the total momentum remains unaltered. If the motion is not confined to one direction in space, the situation is more complicated. We must now select any three directions in space which are mutually at right angles to one another, as South-North, West-East, and down-up. The motion of each particle must now be separated into its constituent motions in these three directions. This of course divides the momentum into three parts, one in each of the three chosen directions. The West-East momentum of the particle is now defined as the mass of the particle multiplied by the speed at which it moves from West to East and so on. It can now be shown that the total momentum in each of the three directions separately must remain unaltered, and the same is of course true of any other direction in space that we may select.

In whatever way a number of particles may move, their motion must always conform to the general principles just stated. If a problem is of a sufficiently simple nature, these principles may suffice of themselves to provide a complete solution, without our troubling about the motions of the individual particles at all.

Suppose, for instance, that in a shunting yard an empty truck weighing 5 tons runs at 5 miles an hour into a loaded truck weighing 20 tons, which is standing at rest. Suppose that the trucks are fitted with an automatic coupling of the American type, so that they become locked together after their impact, and then have to move forward at the same speed; we wish to know what this speed will be.

We need only notice that the forward momentum of the coupled trucks after impact must be exactly equal to their forward momentum before, so that the amount of momentum which originally resided in the one 5-ton truck must now be distributed over 25 tons. This 25 tons will accordingly move forward at one-fifth of the speed at which the 5-ton truck originally moved—the two trucks move forward together at 1 mile an hour.

If there is no automatic coupling, the problem becomes slightly more complicated, because the trucks can then rebound after impact and move at different speeds afterwards. As we have now to find the values of two different quantities—the two speeds after impact—we need two relations from which to find them. A second relation is supplied by the fact that the total kinetic energy must be the same after the impact as it was before. Using these two relations, we find that the loaded truck will now move forward at 2 miles an hour, while the light truck rebounds and moves backwards at 3 miles an hour.

### Equations of Motion

More complicated problems cannot be solved in this simple way, but other and somewhat similar methods are available; let us try to illustrate them by the simplest of examples.

In the game of billiards, three balls roll about on a rough surface bounded by resilient cushions; they move as they are impelled by the impact of extraneous objects, the cues. It would be possible to follow out their motion by treating each ball as an infinite number of minute particles, first reckoning out how each particle pulled or pushed its neighbour, and then calculating the resulting motion of the balls as a whole. This, indeed, is what we should have to do, if we were limited to using Newton's laws in the crude form in which they were originally enunciated. But such a problem would be one for Laplace's infinitely patient mathematician, and not for ordinary mortals, whose life is too short; they need other methods.

The position of any ball on the table can be specified by two measurements, namely the distances of its centre from each of two cushions, one on a long and one on a short side of the table. Such measurements are called *coordinates*. Thus the position of all three balls can be specified by mentioning the values of six co-ordinates.

This takes no account of any spins or rotations the balls may have. Now the orientation of any ball can be specified by mentioning the values of three angles, and these may also be regarded as coordinates, although of a slightly different kind. Thus we see that the positions, not only of the balls as a whole but of every particle in the balls, can be specified by the values of fifteen co-ordinates, six of which measure position and nine orientations. If we are further told the rate at which each of these coordinates is increasing, these fifteen new quantities give us a complete knowledge of the motion of every particle in the balls. These thirty quantities specify the state of the three balls completely.

Thus all the knowledge that Laplace's mathematician would demand for a prediction of the whole future motion of the infinite number of particles in the three balls is contained in the values of only thirty quantities—fifteen coordinates and their fifteen rates of change—and all the information he could give us as to the state of the balls at any future instant would be comprised in the values of these same thirty quantities at that future instant.

Short cuts have been found by which we can pass from the values of the thirty quantities at one instant to those at another instant without troubling about the movements of individual particles, and there are similar methods for tracing out the motions of any mechanical system whatever; the rules for doing so appear in mathematical form, and are known as *equations of motion*. Such sets of equations have been given, in various forms, by a number of mathematicians, especially by Maupertuis, Lagrange and Hamilton.

Hamilton's equations are perhaps the most interesting. They occur in pairs, one pair for each coordinate, and the form of each pair is always the same, regardless of whether the coordinate represents a distance, an angle, or something else. This form of equations of motion is described as the *canonical* form.

We can discover something of the inner meaning of these equations by discussing a very simple case—the motion of a particle moving in a straight line. Here we define the momentum of the moving particle as its speed of motion multiplied by its mass; Newton's second law then tells us that this momentum increases at a rate equal to the force acting on the particle. These statements may be put in the form of equations, thus:

$$\text{mass} \times \text{velocity} = \text{momentum},$$
$$\text{rate of increase of momentum} = \text{force}.$$

Now every pair of Hamiltonian equations is simply a generalization of this pair; the first member of the pair tells us the relation between the velocities of bodies (or, more generally, the rates of increase of coordinates) and certain quantities described as momenta, while the second member tells us the rate at which these momenta increase in terms of the forces, these often including what are usually called centrifugal forces. This second equation is thus a generalization of Newton's second law of motion.

## The Classical Mechanics

So far we have imagined all the energy and all the momentum of the world to reside in the motion of material particles. When it does, we can show, from Newton's laws, that the total kinetic energy of any group of particles will retain a constant value throughout all changes in the motion of individual particles, provided only that no forces act on the group of particles from outside. This is the law of *Conservation of Energy* in its simplest form. The same is true of the total momentum in any direction in space. This is the law of *Conservation of Momentum*.

But when gravitation, chemical action, radiation, electricity and magnetism are taken into account, neither the total energy nor the total momentum of the material particles remains constant. We can, for instance, increase the energy of motion of a motor-car, either by letting it run downhill or by burning some of the petrol in the tank. We cannot of course go on doing this indefinitely, since after a time the car will have dropped to sea-level or the tank will be empty. This leads us to picture both the height above sea-level and the petrol in the tank as representing stores of energy upon which we may draw to increase the energy of the car until these stores are exhausted, but no longer.

To make a consistent picture, we have to suppose that energy can be stored in a great variety of forms, as for example in the raised weight of a clock, in the coiled spring of a watch, in the chemicals used in the cells of an electric accumulator, in the coal we burn in our boilers and in the petrol we burn in our cars. By attributing certain specific amounts of energy and momentum to gravitation, chemical energy, electricity, magnetism and radiation, nineteenth-century physics found it possible to define both energy and momentum in such ways that both were conserved, or at least appeared to be. It was found possible to extend the Newtonian mechanics in this and similar ways until it was able to account for a great range of physical phenomena, and hopes were entertained that in time it would explain all—hopes which, as we shall soon see, were not to be fulfilled.

This extension of the Newtonian mechanics is generally described as the 'classical mechanics'. We are only concerned here with such of its features as are of general philosophical interest.

One of these may be mentioned at once; let us again avail ourselves of a specific example.

Suppose we return to the billiard-table we discussed three pages back, and find that it has been made more complicated in our absence. The original table was suited to the illustration of the Newtonian mechanics, the new to the illustration of the more complicated classical mechanics. Someone has put magnets inside the balls and also inside the cushions of the table, has laid electric wires through the bed of the table and installed batteries and switches to create and control electric currents. To describe the state of this system completely, we shall certainly need more than our original fifteen coordinates, but the classical mechanics assures us that some finite number will suffice, and further provides us with equations of motion for the new coordinates.

It is surprising and significant that these new equations of motion are precisely similar in form to the simple canonical equations of the Newtonian mechanics. That is to say, the same sort of symbols occur in the new equations as in the old, and enter in precisely the same way although of course they have different meanings. The new equations accordingly admit of the same sort of general in--

terpretation as the old; in each canonical pair, one equation tells us, as before, that the momentum associated with one coordinate increases at a rate equal to the force which operates to increase this coordinate; the other specifies the rate of change of this coordinate in terms of the various momenta. This similarity of interpretation shows that the classical mechanics is still fundamentally Newtonian in conception; nature can still be pictured as consisting of particles which are pushed and pulled about by forces.

### Action at a Distance

Difficulties occur as soon as we try to picture these pushes and pulls in detail.

When one billiard-ball strikes another and sets it into motion, it is easy to imagine pairs of molecules, one in each ball, pushing each other and so transmitting force from the one ball to the other; Newton's concept of force makes it possible to form a perfectly definite picture of what happens in such a case. But it is not so easy to picture what happens when the moon raises tides on the ocean, or the sun holds the earth in its orbit. Newton's law of gravitation specified the *amount* of the force acting between two bodies such as the sun and the earth, but made no attempt to explain the *nature* of the force, or how the force could operate across stretches of apparently empty space. How can the moon move the waters of our oceans unless there is some chain of continuous contact between moon and earth—such as might be provided, for instance, by a sheaf of strings or elastics, or perhaps by a liquid transmitting a continuous pressure or tension? What, it may be asked, plays the part in reality of such a system of strings, elastics or liquids?

Newton and his contemporaries asked such questions as these, and it was generally felt that an answer must be found before Newton's theory of gravitation could be accepted. In a famous letter to Bentley, Newton wrote: 'It is inconceivable that inanimate brute matter should, without the mediation of something else which is not material, operate upon and affect other matter without mutual contact.... That gravity should be innate, inherent and essential to matter, so that one body may act upon another at a distance, through a vacuum, without the mediation of anything else by and through which their action may be conveyed from one to another, is to me so great an absurdity that I believe no man, who has in philosophical matters a competent faculty of thinking, can ever fall into it.'

The question remained unanswered until Einstein's generalized theory of relativity came in 1915 and showed that there is probably neither any answer to it, nor need for an answer.

We have already seen (p. 63) that a three-dimensional space does not of itself provide a suitable framework against which to represent the motions of objects. When a number of objects stand at rest, their spatial relations may be represented in a three-dimensional continuum, and such an arrangement, if properly made, will be consistent with itself and will 'make sense'—we shall be able to represent not only some but all of the spatial relations of the objects in a single arrangement. But such an arrangement is found to be inadequate when the objects are in rapid motion; no such arrangement can then represent all the observable facts. A fourth dimension, of the general nature of time, must be added to the three dimensions of simple space, forming the four-dimensional continuum that we have described as the space-time unity (p. 63). We cannot say that any one particular dimension in this represents time, while the other three represent space. The four-dimensional continuum forms an indissoluble unity, and must always be regarded as a whole. In it are any number of different directions, any one of which may be taken to represent time, and will adequately represent it to an observer who is moving through space at the right speed.

This four-dimensional continuum, formed by the indissoluble welding of space and time to form something different from either, is found to provide by far the most suitable framework for the discussion and explanation of the phenomenon of gravitation. A point in the continuum represents *a point of space at an instant of time*. Thus the fact that a gravitating mass such as the sun occupied a particular point of space at a particular instant of time can be represented by the position of one single point *P* in the continuum, while the position of the same mass in space at other instants will be represented by the positions of other points $Q, R, S, \ldots$ in the continuum. The line obtained by joining up the points $PQRS \ldots$ constitutes a record of the various positions occupied by the mass through an interval of time, or through the whole of time if we wish. Such a line is called the 'world-line' of the mass in question.

With this framework before us, we find that a concise, complete and perfect picture of the pattern of events can be obtained in the following way.

We first suppose that the presence of a gravitating mass at the place and time represented by the point $P$ of the continuum impresses a curvature on the continuum in the proximity of the point $P$, just as the presence of a lead ball on a cushion at a certain place and time impresses a curvature on the cushion in the proximity of these points of space and time. Thus the continued existence of the sun will impress a curvature on the continuum in the region surrounding the world-line of the sun.

Having introduced us to a curved continuum in this way, the theory of relativity now tells us that the world-lines of small bodies moving in the neighbourhood of the mass—as, for example, planets, comets or meteorites moving round the sun—are either straight lines, or are the straightest lines that are consistent with the curvature of the continuum.

This simple statement describes the whole pattern of events, except that it must obviously be put in a slightly different form when more than one gravitating mass is involved. If there are no gravitating masses present, the continuum can have no curvature*. Thus the world-line of a particle is a straight line—i.e. the particle continually moves in the same direction and at the same speed. This is Newton's first law, which now appears as a simple inference from the theory of relativity. If gravitating masses are present, a particle appears to move in a curved path, but the apparent curvature of path merely reflects the curvature of the continuum. Newton thought that a planet followed a curved path in a straight (flat) space; the theory of relativity pictures it as following a straight path in a curved space.

* We need not discuss the possibility of space having inherent in itself a curvature on a universe-wide scale; such curvature, if it exists, is unimportant for our present discussion.

We notice that all reference to force has disappeared, so that the motions of the planets and of other gravitating bodies present problems in geometry, but not in dynamics. Also the question of action at a distance has dropped out altogether. Nature has dodged it by the simple manœuvre of making gravitation act on space instead of across or through space, although, in a sense this only postpones the difficulty; it provides a new description, but not a satisfying explanation, of the facts.

At the same time, the question of causality has assumed a new aspect. We can no longer say that the past creates the present; past and present no longer have any objective meanings, since the four-dimensional continuum can no longer be sharply divided into past, present and future. All we can say is that the world-lines of all objects in the universe follow the simple pattern already described. If these world-lines have a real existence in a real continuum, the whole history of the universe, future as well as past, is already irrevocably fixed. If on the other hand the world-lines are merely constructions we draw for ourselves, to help us visualize the pattern of events, then it is as easy to extend these world-lines from our already completed past into our future as it is to carry on the weaving of a fabric when the pattern is already set in the loom. In either case the future is unalterable, and inescapable determinism reigns.

## Electric and Magnetic Forces

Superficially at least the forces of electricity and magnetism seem to present the same kind of problem as the forces of gravitation. Experiment shows that two electrically charged bodies attract one another (or repel if their charges are of the same kind) with a force which conforms to the same mathematical law as the force of gravitation—both forces fall off inversely as the inverse square of the distance. The same is true of magnetic force also; two magnetic poles attract or repel one another with a force which again follows the law of the inverse square of the distance.

This being so, we might well expect these forces to admit of a pictorial interpretation of the same kind as that for gravitational force. But no such interpretation has yet been found, and the prospect of finding one now looks very remote. Electric and magnetic forces in general present a far more intricate problem than gravitational forces. Gravitational force is simple, and a thing by itself, as also are electric and magnetic forces so long as the electric charges and magnetic poles stand at rest. But as soon as motion comes into the picture, the whole situation is changed. Forces of new kinds come into play, for moving electric charges exert magnetic forces in addition to the electric forces they exert when at rest, while moving magnets exert electric forces in addition to the magnetic forces they exert when at rest. When the exact laws governing these intricate laws had been discovered by a great number of experimenters, Clerk Maxwell succeeded in expressing them in a mathematical form which was both simple and elegant.

At this time, space was supposed to be filled with an ether, a substance which might well serve, among other functions, to transmit forces across space. So long as such an ether could be called on, the transmission of force to a distance was easy to understand; it was like ringing a distant bell by pulling a bell-rope.

The pattern of electrical events being known with complete precision in mathematical terms, it was natural to try to discover the properties of the ether from this pattern. It was taken for granted that these properties would prove to be mechanical—either the particles of the ether would be found capable of motion in the Newtonian sense and in accordance with the Newtonian laws, or else they would conform to some more general principle, such as 'least action' (p. 187), which formed a sort of generalization of the Newtonian laws; they would in either case be pushed or pulled about by forces. Faraday, Maxwell, Larmor and a great number of others all tried to explain electromagnetic action on these lines, but all their attempts failed, and it began to seem impossible that any properties of the ether could explain the observed pattern of events.

Then the theory of relativity came and explained the causes of failure. Electric action requires time to travel from one point of space to another, the simplest instance of this being the finite speed of travel of light (p. 63). Thus electromagnetic action may be said to travel through space and time jointly. But by filling space and space alone with an ether, the pictorial representations had all presupposed a clear-cut distinction between space and time. Clearly, if such a distinction existed, it ought to be possible to separate the two out by experiment. Yet when the experiment was attempted by Michelson and Morley it failed, thus showing that the space and time assumed in the picture were not true to the facts of nature.

On this failure the theory of relativity was built. It provided a clue to the solution of the puzzle by showing that the pattern of events could not be altered by making the whole electric structure move through the supposed ether at any speed whatever. This, indeed, was the fundamental postulate of the theory, which every experiment so far made has confirmed—the pattern of events cannot be altered by altering the speed of motion. In other words, the pattern of events was the same whether the world stood at rest in the supposed ether, or had an ether wind blowing through it at a million miles an hour. It began to look as though the supposed ether was not very important in the scheme of things, and further discussion showed that it could not serve any useful purpose and so might as well be abandoned. But if the bell-rope has to be discarded, what is to ring the bell?

Clearly, if electric action is to be explained in mechanical terms, the mechanism must be supposed to be attached to the electric charges, and to move through space with them. It must extend through the whole of space, because the attraction and repulsion of an electron extend through the whole of space, and it must be the same for all directions in space, since an electron at rest exerts a force which is the same for all directions in space. Further, as the pattern of events is unaltered by motion, the mechanism must be the same when the electron is in motion as when it is at rest. But experiment shows that an electron in motion exerts additional forces which are not the same for all directions in space; if we picture this electron as moving head-foremost through space, these forces surround it like a belt round its waist.

Thus direct experimental evidence shows that the forces exerted by an electron (or of course by any other charged body) can neither be attributed to any mechanism attached to the body, nor to action transmitted through an ether or any medium surrounding the body. We have a perfect specification of the pattern of events written, as it necessarily must be, in the language of mathematics, but this does not admit of interpretation in mechanical terms, or indeed in any terms other than those of mathematics.

This is true also of the greater part of the classical mechanics. The only part that we understand pictorially is the Newtonian part which deals with mechanical phenomena on the man-sized scale; we can understand this because the phenomena directly affect our senses; the pictorial explanation is in terms of forces such as we exert with the muscles of our bodies, and the idea of such forces is familiar to our minds.

If we wish to visualize other processes pictorially, no single perfect picture is available, and the best we can do is to construct a number of imperfect pictures, each representing one, but only one, aspect of the complete range of phenomena. For instance, if a shower of electrons is shot on to a zinc sulphide screen, a number of flashes are produced—one for each electron—and we may picture the electrons as bullet-like projectiles hitting a target. But if the same shower is made to pass near to a suspended magnet, this is found to be deflected as the electrons go by. The electrons may now be pictured as octopus-like structures with tentacles or 'tubes of force' sticking out from it in every direction.

It would, however, be wrong to think of an electron as a bullet-like structure with tentacles sticking out from its surface. We can calculate the mass of the bullet, and also the mass of the tentacles. The two masses are found to be identical, each agreeing with the known mass of the electron. Thus we cannot take the electron to be bullet *plus* tentacles—this would give us twice too great a mass—we must take it to be bullet *or* tentacles. The two pictures do not depict two different parts of the electron, but two different aspects of the electron. They are not additive but alternative; as one comes into play, the other must disappear.

Actually the situation is even more complicated, since a separate tentacle picture is needed for each speed of motion of the electron, the speed being measured relative to the suspended magnet or other object on which the moving electron is to act. The reason is that already explained. When the electron is at rest, the tentacles stick out equally in all directions. But an electron which is at rest relative to one magnet may be in motion relative to another, and to discuss the action of the electron on this second magnet we must picture it as having a belt of tentacles round its waist. This shows that we must have a different picture for every speed of relative motion, so that the total number of pictures is infinite, and we cannot form the picture we need until we know the speed of the electron relative to the object it is about to meet.

## THE FAILURES OF THE CLASSICAL MECHANICS

By the end of the nineteenth century the classical mechanics might almost be said to have met with complete success in explaining and predicting the phenomena of what we have called the man-sized world. It had also been very successful with the still larger scaled problems of astronomy, although missing complete success in a comparatively small group of problems which are now, we hope, in process of being cleared up by the gravitational theory of relativity. But at the other end of the scale there was no success at all; experimental physics was particularly interested in the processes taking place inside the atom, and in this field the classical mechanics was failing conspicuously and completely. Perhaps its most spectacular failure was with the fundamental problem of the structure of the atom.

### Atomic Structure

Experimental physics had provided strong reasons for thinking that an atom consists of a collection of electrons—negatively charged particles—together with something which carries just enough positive electricity to counteract the total negative charge of them all—for the total charge on a normal atom is always zero.

Now there is no mechanism within the framework of the classical mechanics for endowing such a structure with a permanent unchanging size. Its charges cannot stand at rest, or they begin to fall into one another, and they cannot be in motion or they become a perpetual-motion machine of the kind not permitted by the classical mechanics. Thus the mere permanence of the atom showed the need for a revision of the classical mechanics.

And whatever system of mechanics we finally adopt, we should expect that the fixed and unchanging sizes of atoms could be calculated by combining the known constants of nature in some way or other. But the constants known to the classical mechanics cannot be combined to form a length at all, and this seemed to suggest that some other fundamental constant of nature still remained to be discovered.

## The Problem of Radiation

Another conspicuous failure of the classical mechanics was with one aspect of the problem of radiation. Here it predicted very general and particularly clear-cut results, which observation was found to negative completely. A simple illustration will explain the nature of the conflict.

Imagine a crowd of steel balls set rolling about on a steel floor. If two balls bump into one another, their individual speeds and directions of motion will change, but the incident will not alter the total energy of motion of the balls. There must, however, be a steady leakage of energy from other causes, such as air resistance and the friction of the floor, so that the balls continually lose energy and, after no great length of time, will be found standing at rest on the floor. The energy of their motion seems to have been lost, although we know that actually most of it has been transformed into heat. The classical mechanics predicts that this must happen; it shows that all energy of motion, except possibly a minute fraction of the whole, must be transformed into heat whenever such a transformation is physically possible. It is because of this that perpetual-motion machines are a practical impossibility.

Precisely similar ideas are applicable to the molecules which form the air of a room. These also move about independently, and frequently bump into one another. The classical mechanics now predicts that the whole energy of motion will be changed into radiation, so that the molecules will shortly be found lying at rest on the floor—as the steel balls were. In actual fact they continue to move with undiminished energy, forming a perpetual-motion machine in defiance of the classical mechanics.

Why does the classical mechanics meet with such different degrees of success in these two cases? Why does it fail so conspicuously for molecules of air, when it gave the right results for steel balls? The short answer is that we have passed from one to another of the three worlds we discussed on p. 42—from the man-sized world to the world of the electron.

We can go further than this. It seems fairly clear—although no absolutely compelling proof can be provided—that if any system of bodies whatever is moving continuously in time and space under any system of laws whatever, provided only that there is a causal law so that one state is followed uniquely by another, then the final upshot of the motion must be that predicted by the classical mechanics —all the energy of the bodies must be transferred from matter to radiation. This fallacious result is not, then, a peculiarity of the classical mechanics; it is given also by a very wide class of possible systems of mechanics. This being so, no minor modification of the classical mechanics can possibly put things right. Something far more drastic will be needed; we are called upon to surrender either the continuity or the causality of the classical mechanics, or else the possibility of representing changes by motions in time and space.

### Motions in Time and Space

Now these three concepts formed the foundation-stones of the philosophy of materialism and determinism to which the physics of the nineteenth century seemed to lead. Thus as soon as any one of the three has to be rejected, the philosophical implications of physics undergo a great change; the mechanical age has passed, both in physics and philosophy, and materialism and determinism again become open questions—at least until we have seen what the new physics has to say about them. We shall discuss this new physics in the next two chapters, and its philosophical implications in our final Chapter VII.

# CHAPTER V

## THE NEW PHYSICS

### (PLANCK, RUTHERFORD, BOHR)

#### PRELIMINARY SURVEY

With the coming of the twentieth century, there came into being a new physics which was especially concerned with phenomena on the atomic and sub-atomic scale. It brought with it a new way of interpreting the phenomena of inanimate nature, which was destined in time to sweep away all the difficulties besetting the old classical mechanics. A preliminary glance over the vast territory of this new physics reveals three outstanding landmarks.

First we notice an investigation which Prof. Planck of Berlin published in 1899. His aim was so to amend the classical mechanics that it should fit the observed facts of radiation, and show why the energy of bodies was not wholly transformed into radiation. We have already seen that this was likely to involve giving up either continuity or causality or the representation of phenomena as changes taking place in space and time. Actually his investigation seemed to show that continuity had to be given up, suggesting that in the last resort changes in the universe do not consist of continuous motions in space and time, but are in some way discontinuous.

The classical mechanics had envisaged a world constructed of matter and radiation, the matter consisting of atoms and the radiation of waves. Planck's theory called for an atomicity of radiation similar to that which was so well established for matter. It supposed that radiation was not discharged from matter in a steady stream like water from a hose, but rather like lead from a machine-gun; it came off in separate chunks which Planck called *quanta*. This, as we shall shortly see, carried tremendous philosophical consequences with it.

An extension of Planck's ideas, due to Prof. Niels Bohr of

Copenhagen, went on to suggest that, viewed through a microscope of sufficient power (this being far beyond anything attainable in practice), the ultimate particles of matter would be seen to move, not like railway trains running smoothly on tracks, but like kangaroos hopping about in a field.

A second conspicuous landmark in the field of the new physics is the enunciation of the fundamental law of radioactive disintegration by Rutherford and Soddy in 1903. This law was in no sense a consequence or development of Planck's theories; indeed fourteen years were to elapse before any connection was noticed between the two. The new law asserted that the atoms of radioactive substances broke up spontaneously, and not because of any particular conditions or special happenings. This seemed to involve an even more startling break with classical theory than the new laws of Planck; radioactive break-up appeared to be an effect without a cause, and suggested that the ultimate laws of nature were not even causal.

A theoretical investigation which Einstein published in 1917 provides a third conspicuous landmark. It connected up the two great landmarks already mentioned by showing that the disintegration of radioactive substances is governed by the same laws as the jumps of the kangaroo electrons in the theory of Bohr. In fact radioactive atoms were now seen merely to contain a special breed of kangaroos, much more energetic and ferocious than any that had hitherto been encountered.

The laws which governed the spontaneous jumps of kangaroos were shown to be of the simplest; out of any number of kangaroos a certain proportion always jumped within a specified time, and nothing seemed able to change this number. Also, before the jumps took place, there was nothing in the world of phenomena to distinguish those kangaroos that were about to jump from those that were not—neither good nor bad treatment could make a kangaroo jump until it hopped out, apparently of its own accord, to help fill the quota demanded by the statistical law. As discontinuity marched into the world of phenomena through one door, causality walked out through another. We shall see later why this had to be.

## PLANCK'S THEORY OF QUANTA

After this preliminary glance, let us turn to a more detailed survey of the situation. Planck's theory asserted that radiation was as atomic in its structure as matter, but with one essential difference. There are only ninety-two different kinds of atoms of matter—or somewhat more when isotopic differences are taken into account— but there are an infinite number of different kinds of radiation, these being distinguished by the different lengths of their waves. Planck found it necessary to postulate an infinite number of kinds of quanta or atoms of radiation, one for every length of wave. The energy contained in an atom, or quantum, of radiation is large when the wave-length is small, and vice versa. The precise relation is that the energy is equal to $h$ times the frequency of the radiation, this being the number of complete wave-oscillations which occur at any specified point in a second, or again the number of complete waves which pass over the point in a second—the two definitions are equivalent. The factor of proportionality $h$ is found to be a universal constant of nature. It is generally known as Planck's constant, and incidentally has dominated atomic physics since its discovery. We have already seen (p. 125) that some such constant was much needed to give a definite size to the atom; here it was.

### The Photo-electric Effect

Not only was Planck's theory immediately successful with those particular problems of radiation for which it had been especially designed, but further confirmation of its truth was soon forthcoming from entirely different quarters. Much of the evidence had been known for some time, but it needed an Einstein to point out its significance (1905).

The evidence in its simplest form was provided by a phenomenon known as the 'Photo-electric Effect'. When ultra-violet radiation (p. 53) falls on a metal surface, a stream of electrons is found to be ejected from the metal. If radiation is pictured as waves, there is no difficulty in seeing in a general way why this should be; the incidence of the radiation may well shake the electrons about in the atoms of the metal, and under very intense radiation they might break loose altogether—like boats breaking loose from their moorings in a stormy sea. Yet if this were the true explanation, weakening the radiation ought to result in the electrons being ejected with less energy, or perhaps not coming off at all. Actually a weakening of the radiation leaves the energy of each electron unimpaired, although reducing the number of electrons shot off. This number is proportional to the intensity of the radiation, so that even the feeblest stream of radiation produces a minute trickle of electrons in which each individual moves just as vigorously as in the bigger flow produced by more intense radiation; it is as though the radiation was a hail of projectiles, hitting some electrons out of their atoms, but leaving the rest untouched.

Further, when an electron is ejected, the total energy it has absorbed from the radiation is found always to be exactly one whole quantum of the radiation. Not all of this energy will appear as energy of motion, since the electron must expend some of it in breaking loose from its atom, and more in fighting its way out through the other atoms to outer space.

We have seen that radiation of low frequency has quanta of low energy and conversely. Radiation may be of such low frequency that the absorption of a quantum by an atom will not liberate an electron; the limiting frequency at which this change occurs is called the *threshold-frequency*. Thus radiation only liberates electrons when its frequency is above the threshold-frequency.

As the amount of energy required to set an electron free naturally depends on the properties of the atom to which the electron belongs, different substances have different threshold-frequencies. Those of most substances are well above the frequencies of visible light, so that the quanta of sunlight and of ordinary room lighting are too feeble to tear electrons off common objects. Even so, they may still carry enough energy to cause some rearrangement of the molecules of the substance on which they fall. Such rearrangement is known as *photo-chemical* action, and it is found that the absorption of a single quantum never affects more than one molecule—this is known as Einstein's law of photo-chemical action. This chemical action of photons explains why bright sunlight causes our curtains and furnishings to fade, and why certain chemicals such as peroxide of hydrogen must be kept away from bright light if the molecules are not to change their composition. It explains too why blue and violet lights—the lights of highest frequencies—affect photographic plates more than lights of other colours.

When the frequency of radiation is above the threshold-frequency, electrons are torn off, and the energy of their motion obviously ought to be proportional to the excess of the radiation-frequency above the threshold-frequency; experiment confirms this law completely.

The process we have been considering is the transfer of energy by radiation from matter at one place to matter at another place; the experiments just mentioned show that this transfer always takes place by complete quanta. Confirmation is provided by Heisenberg's contributions to the subject, which are discussed in the next chapter. Heisenberg finds that facts of observation lead uniquely and inevitably to the theoretical structure known as matrix mechanics. This shows that the total radiation in any region of empty space can change only by a single complete quantum at a time. Thus not only in the photo-electric phenomenon, but in all other transfers of energy through space, energy is always transferred by complete quanta; fractions of a quantum can never occur. This brings atomicity into our picture of radiation just as definitely as the discovery of the electron and its standard charge brought atomicity into our picture of matter and of electricity.

## The Atomicity of Radiation

In 1905 Einstein proposed a pictorial representation of all this, which was in many ways reminiscent of the corpuscular theory by which Newton had tried to explain light two centuries earlier.

Planck had supposed that an atom could only emit radiation by complete units or quanta. Einstein now pictured each emitted quantum as travelling through space in the form of a compact and indivisible unit—an unbreakable packet of radiation. Such a packet he called a *light-arrow*, although the more non-committal term *photon* is more usual to-day.

According to this picture, a stream of radiation may be visualized as a shower of photons. When this falls on a material surface, like a hail of arrows hitting a target, each photon will hit one of the electrons in the surface, and will do damage which is confined to the point of impact. This picture explains at once why weakening the stream of radiation does not stop electrons coming off, why doubling the intensity of radiation doubles the number of electrons and, more generally, why the two are proportional.

Simple considerations of a general kind show that a free electron —i.e. one which is not attached to an atom—can never absorb a quantum of radiation. If, then, a light-arrow should strike such an electron, we must picture the two as colliding like two billiard-balls, and the collision will change the directions of motion of both. In 1925 Compton and Simon were able to photograph the paths of electrons both before and after such 'collisions', and found that the light-arrows of Einstein's picture must be supposed to carry precisely the amounts of energy and momentum that the quantum theory demanded.

### The Undulatory Nature of Radiation

While there is convincing experimental evidence that radiation is both emitted and absorbed in complete quanta, there is none to show that these quanta travel through space in the unbreakable units suggested by Einstein's picture. Indeed, there cannot be; it is only at the beginning of its journey, when it is emitted by matter, and again at its end, when it again interacts with matter, that radiation can make its presence known to our senses or apparatus.

But there is a great deal of evidence that light does not travel through space in the form of these unbreakable units; there is in fact the evidence of the whole undulatory theory of light. It will be enough to illustrate this by a single example, which shows the evidence in a particularly clear form.

Suppose that light of pure colour, and so of uniform wave-length, is emitted from the source of light $S$ in fig. 1. Let us further imagine a screen $AB$, punctured by two movable pinholes at $A$ and $B$, to be set up as shown, and let a second screen be placed behind it, the lines $SA$ and $SB$ meeting this second screen at the points $P$ and $Q$.

When the source $S$ emits light, we should expect to find the points $P$ and $Q$ illuminated, while the rest of the screen remained dark. And so we do, so long as we do not examine the screen too closely; at a cursory glance we might well think that photons had passed like arrows through the holes $A$ and $B$. But a more careful examination shows that the illumination at $P$ and $Q$ consists of something more than the small circular patches of light which the

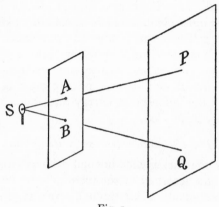

Fig. 1

arrow picture of radiation would lead us to expect; at each of the two points we find a complicated pattern consisting of concentric circles of light alternating with concentric circles of darkness.

Before discussing this, let us extend our experiment by moving the pinholes $A$ and $B$ gradually nearer to one another. At first the patterns at $P$ and $Q$ simply approach one another in the way we should expect, but when they have come quite near to one another a new phenomenon occurs. The pattern we now observe can no longer be obtained by the mere addition of the two circular patterns at $P$ and $Q$. These patterns have begun to interact with one another, and for certain positions of $A$ and $B$, the points $P$ and $Q$ become completely dark. Keeping $A$ and $B$ in such positions, let us stop up the hole $B$. We find that the point $P$ immediately changes from being dark to being light. If we unstop $B$, $P$ becomes dark again. Thus to all appearances a decrease in the illumination *adds to* the light at $P$, while an increase *subtracts from* it.

Such results obviously cannot be explained in terms of photons travelling like arrows through holes. The undulatory theory, on the other hand, explains them at once. It tells us that the illumination at each point is produced by the combined action of two sets of waves, one coming through $A$ and one through $B$, and it is one of the commonplaces of physics that two such sets of waves can neutralize one another. The process consists in the crests of one set of waves coinciding exactly with the troughs of the other set, so that the effects of the two sets of waves just cancel out, and is known as *interference*. This not only provides a general explanation of the phenomenon, but also enables us to predict the pattern completely.

### The Particle- and Wave-pictures

We now have two distinct pictures of the nature of radiation, one depicting it as particles and the other as waves. The particle-picture is obviously the more suitable when the radiation is falling on matter, and the wave-picture when it is travelling through space. For a time there was a disposition to conclude that light must consist of two parts, a wave part and a particle part, but it is now clear that this is not so. The wave-picture and the particle-picture do not show two different things, but two aspects of the same thing. They are simply partial pictures which are appropriate to different sets of circumstances—like the two pictures of the electron which we introduced on p. 122—and so are complementary but not additive. As soon as light shows the properties of particles, its wave properties disappear, and vice versa; the two sets of properties are never in evidence at the same time. Thus as we follow a beam of light, or even a single quantum, in its course, we must imagine the wave- and particle-pictures taking control of the situation alternately.

The wave-picture explains much in its own proper province,

but it brings its own difficulties with it. In particular, it is not easy to pass back from the wave-picture to the particle-picture. For all waves scatter as they travel through space, and it is difficult to imagine how waves which have once scattered as the undulatory theory directs can recombine and concentrate their attack on single molecules or electrons in the way they are observed to do as soon as they encounter matter.

Suppose, for instance, that the source $S$ (p. 132) emits only a single quantum of light. If this travels through space in the form of the waves required by the undulatory theory, some of it must pass through the hole $A$ and some through the hole $B$, while the greater part will be absorbed or reflected by the screen $AB$. We cannot imagine all these various parts recombining and directing the whole of their energy upon a single molecule, either on the near or on the far side of the screen $AB$, so that our picture seems to fail entirely. We must always remember that the actual physical processes are essentially unpicturable, but obviously their results cannot be obtained from any activities which we can imagine operating in time and space, so that we here obtain our first intimation that the space-time framework of the classical mechanics is inadequate for the complete representation of natural phenomena.

The undulatory theory of light attained its most precise, and (as many then thought) its final, form in Maxwell's electromagnetic theory of light. This interpreted the waves of the undulatory theory as oscillating electric and magnetic forces travelling through an ether. At each instant of time there was at every point of this ether a definite electric force (which Maxwell tried to represent as a 'displacement' of the ether), and a definite magnetic force—just as, at any point on the surface of a stormy sea there is a definite elevation above, or depression below, the mean level of the sea.

With the passing of absolute space, these ideas became untenable. The theory of relativity washed away the ether, and not only showed that different observers might assign different measures to the forces at the same point and at the same instant of time, but also that they could all be equally right. The so-called electric and magnetic forces, then, are not physical realities, as, for instance, displacements of the ether would have been; they are not even objective, but are subjective mental constructs which we have made for ourselves in our efforts to interpret the waves of the undulatory theory. Indeed, as they were created in an attempt to provide a mechanical explanation of the propagation of light, they come under the same condemnation as the electric and magnetic forces with which we tried to explain the action of an electric charge (p. 119)—and, *mutatis mutandis*, for the same reasons. Clearly we must search for a better interpretation of the waves of the undulatory theory.

### Waves of Probability

Let us return to the imaginary experiment of p. 134, in which a single quantum of radiation is emitted from a source of light to fall on one point or another of a system of distant screens. We know that the whole energy of the quantum will concentrate on a single point of the screens, but which point will it be?

The obvious answer is that sometimes it will be one point, and sometimes another. It cannot always be the same point, or else when quanta were being emitted in millions, this one specially favoured point would be intensely bright and all others completely dark. Actually when quanta are being emitted in millions, there are some places on the screens at which the illumination is very bright, these indicating regions in which many photons have struck, and also places of less illumination, these indicating regions in which few photons have struck. Even the most faintly illuminated parts of the screen must have been struck by some photons.

If we now fix our attention on a single quantum of radiation of which we know nothing except that it belonged to the original beam, we may say that the extent to which either screen is illuminated at any point gives a measure of the *probability* that the quantum shall condense into a photon at this point. In this way we may interpret the waves of the undulatory theory as waves of probability; the extension of the wave system in space marks out the region within which a photon may be supposed to be travelling, while the intensity of the waves at each point within this region gives a measure of the probability that a photon will occur at that point if matter is placed there.

When half a million babies are born in England in a year, we may say that 20 per cent of them are born in London, 2 per cent in Manchester, 1 per cent in Bristol, and so on. But when we think of the one baby born in a single minute of time, we cannot say that 20 per cent of it is born in London, 2 per cent in Manchester, and so on. We can only say that there is a 20 per cent probability of its being born in London, a 2 per cent probability of its being born in Manchester, and so on. If we disregard variations of birth-rate with locality, a map exhibiting the density of population in different parts of England will also act as a chart showing the number of births per annum; but with reference to the birth occurring at any one instant, it merely shows the relative probabilities of the baby appearing in different areas. As soon as the waves of the undulatory theory fall on matter, they provide a precisely similar chart for the probability of photons appearing in the different areas of the matter. The waves, then, are again mental constructs—not enabling us to see what *will* happen, but what *may* happen.

## Waves of Knowledge

The waves may equally well be interpreted as representations of our knowledge. In the experiment with the single photon, we do not know where the photon is, but the wave-picture gives a sort of diagrammatic representation of what we do know. We know that the photon must be within a certain region of space, this being the region mapped out by the waves at each instant. We may know that it is much more likely to be in a region $A$ than in some other region $B$; if so, the waves represent this knowledge by being much more intense in the region $A$ than in the region $B$, and so on.

These two interpretations of waves—as representations of probability and of knowledge—are well illustrated in an idealized experiment imagined by Einstein and Ehrenfest.

An ordinary glass mirror functions because a thin coat of silvering on its back reflects all light falling on it. The silvering can be made so thin that the mirror will reflect only a part of the light falling on it—for simplicity let us suppose half—while the rest goes through to the further side and continues on its way as though the mirror had not been in its path. When a beam of radiation falls on such a mirror, we must imagine that half of its quanta are reflected and half transmitted.

But suppose that only a single quantum falls on the mirror. As quanta are indivisible, we must picture the whole of the radiation as going either one way or the other; the most we can say is that there is a 50 per cent chance that it will be reflected, and a 50 per cent chance that it will be transmitted.

So far the waves have been figuring as representations of probability, telling us the relative probabilities of the quantum being in one or other of the two paths. Suppose, however, that we now place a screen across the path of reflection, and allow a solitary quantum to fall on the mirror as before. If the quantum happens to be reflected, the screen will be struck by a photon, and we can detect its presence (in principle) in a variety of ways, mechanical or photographic. If the photon shows itself in the path of reflection, the intensity of the waves in the transmitted beam is immediately reduced to zero. We may either say that this is because the probability that the photon is following this path has been reduced to zero, or that it is because we now know that the photon is not on this path. If, on the other hand, no photon is seen to strike the screen, the transmitted beam is immediately doubled in strength, while the reflected beam is annihilated, and the same two interpretations are available as before. It may seem odd that we can annihilate a beam of light by conducting an experiment at an unlimited distance away, but this becomes obvious when we consider that the beam is a representation of our knowledge, so that if our knowledge changes abruptly, the beam must also change abruptly. A simple analogy may clear up the matter and show that there is nothing mysterious or mystical about it.

Imagine a ship crossing the Atlantic from New York to Southampton. The first day out, the ship's position would normally be determined by taking the altitude of the sun at noon; the navigating officer would then mark this position on the ship's chart. If the sky was too cloudy for the sun to be seen, it might be necessary to calculate an approximate position by dead reckoning; the officer would know the approximate speed of the ship, or the distance it had travelled through the water as recorded by the log, and could make a rough allowance for the motion superposed by currents in the sea. He might in this way be able to fix his position to within, say, 5 miles. He could not mark a cross on his chart to fix his position, but might draw a circle 5 miles in diameter; this, like the waves of the undulatory theory, would represent his knowledge of his position. As the ship progressed on its journey, we can picture this circle travelling over the chart, like a wave travelling through space, at a speed representing the speed of the ship. As new uncertainties accumulated, the circle would continually increase in size. If the sun was still invisible on the next day, it might be necessary to indicate the ship's position by a circle 10 miles in diameter. If the sun could not be seen throughout the voyage, the uncertainty as to the ship's position would continue to increase, until, by the time the ship was close to land, it might have to be represented by a circle 50 miles in diameter. Suppose, that when such a circle had been marked on the chart, half of it was found to lie over the Cornish coast. As the ship could not be on land, this half of the circle could at once be ruled out; this bit of knowledge would at once reduce the extent of the uncertainty to half— just as happened in the experiment with the half-silvered mirror. If the Lizard was sighted a few moments later, the further knowledge thus provided would reduce the uncertainty practically to zero, and the ship's position could now be marked by a point.

This analogy clears up the physical situation in other respects. We know how in practical life one uncertainty leads to another; for instance, the uncertainty which prevailed as to the ship's position when it was one day out continually increased; this uncertainty made it impossible to allow exactly for the currents encountered on the second day's run, and as the voyage proceeded uncertainty was piled on uncertainty. The wave-picture of radiation faithfully reproduces this cumulative property of uncertainty in knowledge, because it is an inherent property of a group of waves always to spread out, and so occupy more space.

In this analogy the ship represents a photon, the sea represents the space in which the photon moves, and the land represents barriers, such as the screen on p. 132, which prevents the photon moving through the whole of space. The sea, land, ship and photons all exist and move in the ordinary space of everyday life; indeed this is what we mean by ordinary space—the space in which we see things through the impact of photons on our retina, and travel by ship. But the waves which represent the navigator's knowledge of his ship's position do not travel through ordinary space, but over a nautical chart, which is a sort of diagrammatical representation of ordinary space. In precisely the same way, the space traversed by those waves which represent our knowledge of photons is not ordinary space but a mathematical representation of ordinary space. If it contains barriers, these are representations of barriers in ordinary space—like the coastline on a nautical chart. In brief, the space of photons is ordinary physical space, while the space traversed by the waves of the undulatory theory is a conceptual space. Indeed, it must be, since the waves, as we have seen, are mere mental constructs and possess no physical existence.

It may be thought that if we are concerned only with mathematical representations, it is a matter of indifference whether we imagine them set up in ordinary space or in some conceptual space of our own construction. This is so, provided the two spaces have the same number of dimensions. And, as the waves of the undulatory theory of light need a conceptual space of three dimensions for their representation, generations of physicists have identified this with ordinary physical space, and thought of light as waves travelling through the space of everyday life in which we travel by car or train. This is now seen to be a little irrational—rather like marking out the time-table of a railway along the tracks. It can, however, find justification in the fact that an ordinary beam of light contains so many photons that probabilities may be replaced by actualities. When we take this step, the space through which probabilities of photons travel becomes identical with the space through which the photons themselves travel, and this is the space of everyday life—the space in which we see things. In this way we come back to the view of light propagation which all physicists held as a matter of course before the quantum theory came to trouble them.

## The Uniformity of Nature

Before the quantum theory appeared, the principle of the uniformity of nature—that like causes produce like effects—had been accepted as a universal and indisputable fact of science. As soon as the atomicity of radiation became established, this principle had to be discarded.

In the experiment described on p. 135, the uniformity of nature would have required that every photon should hit the screen at the same point. Actually we have seen that they hit at different points, so that if a single quantum is discharged from the source several times in succession, different experiments will be found to give different results, and this although the conditions before the experiments were, so far as we could tell, precisely identical.

The same thing is shown, even more convincingly, by the experiment with the half-silvered mirror. If we shoot solitary photons, one after another at the same point of the mirror, half of them will get through and half will not, so that again a succession of similar experiments will not give similar results.

It may perhaps be objected that if the results of two experiments differ, this must be because either the conditions before the experiments, or else the conditions during the experiments, were not absolutely identical. If we shoot peas at a piece of wire-netting, we may find that half of them get through, while half of them hit the wires of the netting and fall back. If we only shoot a single pea, there is a 50 per cent chance that it will get through. If we shoot a second pea, aiming it so that it meets the netting at precisely the same point as the first, and so making the conditions of the experiments absolutely identical, we may be sure that the experiments will have the same result; if the first pea gets through, the second will also get through. If the two peas were observed to meet different fates, we should conclude that the conditions of the two experiments had not been absolutely identical. It may be objected that similar considerations apply also to the experiments just described, and that if the two quanta of radiation had different experiences, the conditions of the experiments cannot have been absolutely identical.

The conditions of the two sets of experiments are, however, not parallel. In the experiment with the wire-netting, all those peas which failed to get through, as well as many of those that got through, hit the wires of the netting at some point or other, and the exact point at which they hit decided both their fate in the matter of getting through, and also the angle at which their paths lay after impact. Both sets of peas moved at all sorts of angles. But in the experiment with the mirror, all the radiation which gets through moves along exactly the same path, and the same is true of that which is reflected. It follows that the angles of these paths are not determined by the positions of individual molecules, but by the direction of the surface as a whole, and this is sufficient to show that the phenomenon is not molecular or atomic.

In this way we find that the atomicity of radiation destroys the principle of the uniformity of nature, and the phenomena of nature are no longer governed by a causal law—or at least if they are so governed, the causes lie beyond the series of phenomena as known to us. If, then, we wish to picture the happenings of nature as still governed by causal laws, we must suppose that there is a sub-stratum, lying beyond the phenomena and so also beyond our access, in which the happenings in the phenomenal world are somehow determined.

It is natural to wonder why the atomicity of radiation carries more far-reaching consequences than the similar atomicity of matter. But we shall soon see that the atomicity of matter entails precisely similar consequences, the only difference being that these have not been recognized for so long.

## The Principle of Uncertainty

Of the further consequences which follow from the atomicity of radiation, one is of the utmost importance to physics as a whole, and especially to those aspects of it that are under discussion in the present book. Physics sets before itself the task of coordinating the various sense-data which reach us from the world beyond our sense-organs. If our senses could receive and measure infinitely delicate sense-data, we should be able in principle to form a per-fectly precise picture of this outer world. Our senses have limitations of their own, but these can to a large extent be obviated by instrumental aid; telescopes, microscopes, etc. exist to make good the deficiencies of our eyes. But there is a further limitation which no instrumental aid can make good; it arises from the circumstances that we can receive no message from the outer world smaller than that conveyed by the arrival of a complete photon. As these photons are finite chunks of energy, infinite refinement is denied us; we have clumsy tools at best, and these can only make a blurred picture. It is like the picture a child might make by sticking indivisible wafers of colour on to a canvas. We might think we could avoid this complication by using radiation of infinite wave-length. For the quanta of this radiation have zero energy, and so might be expected to provide infinitely sensitive probes with which to explore the outer world. And so they do, so long as we only want to measure energy, but a true picture of the outer world will depend also on the exact measurement of lengths and positions. For this, long-wave quanta are useless. To measure a length accurately to within a millionth of an inch, we must have a measure graduated to millionths of an inch; a yard-stick graduated only to inches is useless. Now quanta of one inch wave-length are, in a sense, graduated only to inches, while quanta of infinite wave-length are not graduated at all. Passing from quanta of short wave-length to quanta of long wave-length only shifts, but does not remove, the difficulty.

A rough analogy is to be found in the problem of photographing a rapidly moving object. A sensitized film can record no detail on a scale which is smaller than the grain of the film, so that if we use a large-grained film, all the fine detail of our picture will be blurred. If we try to escape this difficulty by using a film of very small grain, we merely cross over from Scylla to Charybdis; the speed of the film is now reduced so much that the picture is blurred through its subject having moved appreciably during the time necessary for exposure.

We shall return later to a more detailed discussion of the physical consequences of this. For the moment we pass to yet another consequence of the general fact that our knowledge of the outer world comes to us only through the impact of complete quanta.

## Subject and Object

It used to be supposed that in making an observation on nature, as also in the more general activities of our everyday life, the universe could be supposed divided into two detached and distinct parts, a perceiving subject and a perceived object. Psychology provided an obvious exception, because the perceiver and perceived might be the same; subject and object might be identical, or might at least overlap. But in the exact sciences, and above all in physics, subject and object were supposed to be entirely distinct, so that a description of any selected part of the universe could be prepared which would be entirely independent of the observer as well as of the special circumstances surrounding him.

The theory of relativity (1905) first showed that this cannot be entirely so; the picture which each observer makes of the world is in some degree subjective. Even if the different observers all make their pictures at the same instant of time and from the same point of space, these pictures will be different unless the observers are all moving together at the same speed; then, and then only, they will be identical. Otherwise, the picture depends both on what an observer sees, and on how fast he is moving when he sees it.

The theory of quanta carries us further along the same road. For every observation involves the passage of a complete quantum from the observed object to the observing subject, and a complete quantum constitutes a not negligible coupling between the observer and the observed. We can no longer make a sharp division between the two; to try to do so would involve making an arbitrary decision as to the exact point at which the division should be made. Complete objectivity can only be regained by treating observer and observed as parts of a single system; these must now be supposed to constitute an indivisible whole, which we must now identify with nature, the object of our studies. It now appears that this does not consist of something we perceive, but of our perceptions; it is not the object of the subject-object relation, but the relation itself. But it is only in the small-scale world of atoms and electrons that this new development makes any appreciable difference; our study of the man-sized world can go on as before.

For instance, when an astronomer is observing the motion of a planet in the solar system, it is emitting millions of quanta every second, some of which pass through the telescope of the astronomer and into his eye. By noting the directions from which these arrive, he can follow and describe the motion of the planet across the sky. With the departure of each quantum, the planet suffers a recoil which changes its motion, but the changes are so minute that they may properly be disregarded. But it is different when a physicist tries to follow the motion of an electron inside an atom. He can only obtain knowledge of the internal state of the atom by causing it to discharge a full quantum of radiation, and we shall soon see (p. 146) that the emission of a quantum of radiation is so atom-shaking an event that the whole motion of the atom is changed, and the result is practically a new atom. A succession of quanta may give scraps of information about various stages of the atom, but can give no record of continuous change. In fact there can be no continuous change to record, since every departure of a quantum breaks the continuity.

For this reason it is futile to discuss whether the motion of the atom conforms to a causal law or not. The mere formulation of the law of causality presupposes the existence of an isolated objective system which an isolated observer can observe without disturbing it. The question is whether he, noticing that such a system is in a certain state at one instant, can or cannot foretell that it will be in some other specifiable state at some future instant. But if there is no sharp distinction between observer and observed, this becomes meaningless since any observation he makes must influence the future course of the system.

In more general terms, we may say that the law of causality acquires a meaning for us only if we have infinitesimals at our disposal with which to observe the system without disturbing it. When the smallest instruments at our disposal are photons and electrons, the law of causality becomes meaningless for us, except with reference to systems containing immense numbers of photons and electrons. For such systems the classical mechanics has already told us that causality prevails; for other systems causality becomes meaningless so far as our knowledge of the system is concerned; if it controls the pattern of events, we can never know it.

We have now seen that six important consequences follow from the mere fact of the atomicity of radiation, coupled with those well-established facts of the undulatory theory of light that have been mentioned. These are:

(1) So far as the phenomena are concerned, the uniformity of nature disappears.

(2) Precise knowledge of the outer world becomes impossible for us.

(3) The processes of nature cannot be adequately represented within a framework of space and time.

(4) The division between subject and object is no longer definite or precise; complete precision can only be regained by uniting subject and object into a single whole.

(5) So far as our knowledge is concerned, causality becomes meaningless.

(6) If we still wish to think of the happenings in the phenomenal world as governed by a causal law, we must suppose that these happenings are determined in some substratum of the world which lies beyond the world of phenomena, and so also beyond our access.

### Bohr's Theory of Atomic Spectra

Let us now pass from the general inferences to be drawn from the quantum theory to particular developments of it. Perhaps the most striking of these appeared in 1913, when Bohr suggested that it would provide a solution of the long-standing puzzle of atomic spectra.

In 1911 Rutherford had pictured the atom as a miniature solar system—a crowd of electrons revolving round a massive central nucleus; the electron had to be in orbital motion round the nucleus to escape falling into it. We have already seen (p. 123) that such a picture was incompatible with the classical mechanics; according to this, the electron would continually radiate energy as the result of its orbital motion, and so would gradually spiral down into the nucleus, which would finally absorb it. Thus atoms would be temporary structures of varied and ever-varying sizes.

Bohr planned to remedy these defects by introducing an atomicity of energy into the atom itself. We can explain this sufficiently by considering the simplest kind of atom—the hydrogen atom, in which only one solitary electron revolves round the nucleus. Bohr assumed that the atom could not be of any size whatever, but only of the sizes in which it contained 1, 2, 3, ... or some other integral number of quanta of energy. Hitherto the energy of a quantum had always been $h$ times the frequency of the radiation to which the quantum belonged, but there was now no radiation to provide a measure of frequency. Bohr accordingly measured his quanta against the frequency with which the electron described its orbit.

In this way he avoided both the continual diminution of size of the atom, and its continuous leakage of energy, but there was no opportunity left for the atom to radiate at all. Yet hydrogen atoms certainly could both emit and absorb radiation. He accordingly supposed that the electron did not permanently remain in the same orbit in the atom, but occasionally jumped from one of the permitted orbits to another—these are the kangaroo jumps of which we have already spoken (p. 127); again the process is unpicturable in its ultimate details. Whenever the electron changed its orbit, the intrinsic energy of the atom changed, so that energy was either liberated or absorbed. Bohr supposed that, in either case, the energy thus liberated or absorbed formed precisely one quantum of radiation. This of course fixed the frequency of the radiation. In every previous application of the quantum law, Planck's law, that the energy is $h$ times the frequency, had been used to deduce the energy of a quantum when the frequency of the radiation was already known. In the present case the formula was used the other way; the energy of the emitted photon was known to begin with, and the formula was utilized to deduce its frequency. The frequencies calculated in this way are found to agree completely and exactly with those observed in the spectrum of hydrogen.

This spectrum is of the type known in spectroscopy as a *line-spectrum*. Its appearance is that of a group of bright lines on a dark background, indicating that the radiation divides itself between a number of clearly defined frequencies, and that there is no radiation in between. Before Bohr's explanation appeared, these frequencies had been supposed to belong to some sort of vibration taking place in the hydrogen atom—like the frequencies of the musical note which is heard when a bell or piano-wire is made to vibrate. It now became clear that they had an entirely different origin. The energy exhibited in the spectrum was not liberated by a vibration, or by any kind of continuous motion, but by the sudden jump of an electron to an orbit of lower energy, and its frequency was determined by the compulsion put upon it to form a single quantum.

In the same year in which Bohr produced this revolutionary theory, Franck and Hertz passed a beam of slowly moving electrons through a gas, and measured the amounts of energy that individual electrons yielded up to the molecules of the gas at collisions. The various amounts which the electrons were found to have lost proved always to be one or other of the various amounts needed to raise the atoms from one of the states permitted by Bohr's theory to another. This showed that these states had a real existence, and that transitions between them actually occurred.

To sum up, the success of Bohr's theory suggested that an atom was not a continuously varying structure, from which radiation trickled away like gas from a leaky balloon, but a structure which emitted and absorbed radiation in definite packets at definite instants of time. Thus the energy of the atom did not vary continuously, but jumped suddenly at these instants from one value to another. Only certain definite calculable values were permissible for the energy; these formed a chain of 'energy-levels', arranged like the steps of a ladder, and the energy of an atom could step from any one to any other but could not stand suspended in midair between two steps. When an atom stepped to a lower energy-level, its constituents rearranged themselves suddenly—like the collapsing of a house of cards.

## The Fundamental Laws of Radioactivity

The second great landmark in the new physics is the discovery of the fundamental laws of radioactivity by Rutherford and Soddy.

In 1898 and the immediately succeeding years, Becquerel and the Curies had discovered a group of substances, subsequently known as 'radioactive substances', which possessed very unusual properties, such as a capacity to fog photographic plates kept in their vicinity, and to stand permanently at a slightly higher temperature than the objects surrounding them. In due course the explanation was found; the newly discovered substances not only emitted the normal radiation appropriate to their temperature, but other and additional radiation as well—radioactive radiation as we may call it—from sources which seemed to be internal to the atom. This radiation was finally traced to its origin—or rather origins, for there proved to be three, all of the nature of internal explosions.

Each atom of a radioactive substance can be pictured, like other atoms, as a central nucleus with its crowd of surrounding electrons. The central nucleus must not be pictured as a structureless particle, but as a complex arrangement of many constituents. These constituents, it was found, may suddenly rearrange themselves, and in so doing may eject either a massive particle (known as an α-particle) or a very rapidly moving electron (known as a β-particle) or a quantum of very high frequency radiation (known as a γ-ray).

These three processes may all be included under the common term 'radioactive transformation', since each transforms the original radioactive atom into something different. It was soon found that most radioactive substances had their own characteristic type of radiation, each atom of a substance $A$ transforming into an atom of some other substance $B$, this into an atom of $C$, and so on. Thus, apart from unimportant exceptions, radioactive transformation follows a single one-way track with no branches.

The next step was to investigate the speed with which an atom travels along this path. Ordinary radiation is emitted at a rate which is determined by the temperature of the emitting substance, hot matter emitting radiation profusely and cold matter meagrely. It might not unreasonably have been expected that the same would be true of radioactive emission, but experiment showed that it is not. Two similar masses of radioactive substance may be taken, one of them heated to the highest, and the other cooled to the lowest, of the temperatures available in the laboratory, and both will still emit their radioactive radiation at exactly the same rate as before.

The same is found to be true of all other physical changes. In a milligram of radium, about 500 million atoms disintegrate every second, each giving out its characteristic radiation, and nothing that can be done to the radium or to its environment will change either the number of atoms which disintegrate, or the quality of the resulting radiation. The radiation may be described as spontaneous in the sense that its amount and quality are determined from inside and not from outside.

Such is the fundamental law of all radioactive disintegration, which Rutherford and Soddy enunciated in 1903. It was entirely different in character from any natural law hitherto known, and made it clear that nature proceeded on a plan which was entirely different from anything hitherto suspected.

Interesting but difficult questions arise when we discuss which atoms will disintegrate first, and which will survive longest without disintegration. In the particular instance just given, 500 million atoms are due to disintegrate in the next second. What, we may inquire, determines which particular atoms will fill the quota?

It cannot be anything in the present physical condition or environment of individual atoms, for if it were, we could make more or fewer atoms disintegrate by modifying the physical state of the radium as a whole, and so altering the states of individual atoms. Neither can it be anything in the past histories of the atoms, for if it were, assemblies of atoms with different past histories would show different rates of disintegration, and this again is contrary to the facts;—the rate of disintegration is found to be precisely the same for young atoms of radium which have just been formed by the disintegration of heavier elements as it is for old veterans which are the sole survivors of a stock of radium many thousands of years old. Clearly, then, it is not a case of the young surviving and the old falling. We must rather picture the atoms of radium as drawing lots, young and old on the same footing—like shipwrecked mariners on a raft drawing lots to determine which is to survive. But there is no drawing of lots in nature, so that the choice of one particular atom rather than another appears, from our present point of view, to be an event without a cause.

While the interest of all this to physics was immense and far-reaching, the interest to philosophy was, if possible, even greater,

since it seemed to remove causality from a large part of our picture
of the physical world. We have, let us say, half a million atoms of
radium in this room. If we are told the position and the speed of
motion of every one of them at any moment, we might expect
that Laplace's super-mathematician would be able to predict the
future of every atom. And so he would if their motion had conformed to the classical mechanics. But the new laws merely tell
him that one of his atoms is destined to disintegrate to-day, another
to-morrow, and so on. No amount of calculation will tell him
which atoms will do this; we must rather picture Fate as picking
out her atom, by methods undiscoverable by us. This will then
eject its α-particle, which will proceed to mix with the other atoms
and disorganize their motions—in one way if it is atom $A$ that
disintegrates, but in some quite other way if it is atom $B$. From
the state of the matter at one instant, it is impossible in principle
to discover what the state will be at a future instant.

### Einstein's Synthesis

A third landmark was reached in 1917, when Einstein linked up
these surprising (as they then seemed) laws of radioactive transformation with the equally surprising laws of Planck's quantum
theory.

We have seen how the electrons in an atom can rearrange themselves in new positions of higher or lower energy, and we have
compared falls into positions of lower energy with the collapsing
of a house of cards. A cannon-ball at, say, 1000° F. consists of
atoms of iron, and while the majority of these are in the collapsed
state, some are in states of higher energy, like standing houses of
cards. A wind blowing over a town of card-houses may blow some
down, but it may also cause some which have already been blown
down to stand up again—or so we may imagine for purposes of
illustration. It is much the same inside the cannon-ball. Every
small particle of it is emitting radiation in all directions, and as this
radiation falls on the atoms it may change their condition, causing
some of the standing houses of cards to collapse and some of the
collapsed houses to stand up again. If this were all, it would be
easy to discover how many houses of cards would be standing at

any assigned temperature, how many would be fallen down, and what the constitution of the radiation would be. But the results obtained on this hypothesis do not agree with the facts of observation.

Einstein obtained agreement, brilliantly and completely, by the introduction of a single additional supposition. He supposed that the standing houses of cards could not only be knocked down by the impact of radiation, but that they could sometimes collapse of themselves—in the same way, and according to the same laws, as atomic nuclei collapse in radioactive disintegration, the rate of collapse being entirely independent of environment and physical conditions.

In its new appearance, the law is not concerned with the rather recondite phenomena of radioactivity, but with familiar everyday radiation: it governs the radiation which the sun showers on the earth by day, as well as the light of the electric torch which lights our footsteps at night. Every atom in the universe is not only liable to spontaneous collapse, but also does collapse at frequent intervals. Thus the abdication of determinism appears to be complete, not only from the domain of radioactivity, but from the whole realm of physics.

## Determinism in Nature

Yet science had hitherto been based on the assumption of the uniformity of nature—like causes produce like effects—and if this fails, the whole of science would seem to be left hanging in the air, with no justification for its existence, and no explanation of its success. Yet the success is indisputable, and explanation there must be.

The explanation is twofold. In the first place, the indeterminism disclosed by the quantum theory is confined to the small-scale processes of nature, and in the second place even these indeterminate events are governed by statistical laws. In all man-sized phenomena, billions of electrons and atoms are involved, and for the discussion of such phenomena as are perceptible to us, these may be treated statistically as a crowd. But these crowds obey statistical laws which now take control of the situation, with the result that the phenomena can be predicted with almost the same precision as though the future motion of each particle were known. In the same way, the statistician, knowing the birth-rate, death-rate, etc. of a population, can predict the future changes in the population as a whole, without being able to predict what each separate individual will do in the matter of births and deaths. On the man-sized scale, and indeed far below—down to pieces of matter far too small to be seen in any microscope—nature is, to all appearances, strictly deterministic; like causes produce like effects. Thus the uniformity of nature is re-established except in the realm of the infinitesimal, and science can justify the fundamental assumption on which her existence rests. We see why determinism has become ingrained in our modes of thought, and how Descartes and his followers came to announce it as *a priori* knowledge which they saw by the clear vision of their intellects. Nevertheless, it may not be true for those ranges of nature which were not accessible to them.

# CHAPTER VI

## FROM APPEARANCE TO REALITY

### (BOHR, HEISENBERG, DE BROGLIE, SCHRÖDINGER, DIRAC)

The new physics just described was still based largely on Newtonian ideas. Indeed, in its theoretical aspects, it might not unfairly be described as a final attempt to explain the world in materialistic terms—as particles being pushed and pulled about in space and time. Nevertheless, the new physics had found it necessary to abolish most of the forces of pushing and pulling, replacing the gradual changes of motion of the particles under these forces by sudden and unpredictable jumps. These appeared to involve violations of the law of causality, both in the disintegration of radioactive atoms and also in the internal changes of ordinary atoms. We seemed to see Fate defying this law as she picked out certain atoms for disintegration or collapse and, by her apparently capricious acts, sent the universe along one path or another according to her whim.

On such lines the new physics had explained many phenomena which had hitherto seemed inexplicable, but it had by no means met with complete success. For instance, while it gave a perfect interpretation of the simplest spectrum of all, namely that of the hydrogen atom, it failed with more complex spectra. This was not necessarily a fatal objection; a few emendations and possibly a few new *ad hoc* assumptions might have effected a complete reconciliation, although this seems improbable. What seemed far more serious to many was that success had been achieved only at the price of ejecting continuity and causality from the scheme of nature, and replacing the exact laws of the classical mechanics by an assemblage of statistical laws—and this without disclosing any reason why these statistical laws should be obeyed.

Perhaps this ought not to have been a matter for surprise. We have already seen that the erroneous predictions of the classical mechanics are probably inevitable in any scheme which pictures physical processes as happenings in space and time, and also assumes causality and continuity in these happenings. Planck's original quantum theory attempted to remedy these shortcomings by postulating processes of a very novel kind, but these were still supposed to occur in space and time. This being so, it was almost a foregone conclusion that either causality or continuity would have to be renounced, and there was no special reason for surprise when it was found necessary to renounce both. These general considerations were not, however, widely appreciated at the time, so that few scientists and perhaps even fewer philosophers were prepared to accept the discontinuities and indeterminism of the old quantum theory as final.

## The New Quantum Theory

In 1925 Heisenberg made a new attempt, on entirely novel lines, to obtain an explanation of atomic spectra. Working in collaboration with Bohr, he had come to the conclusion that the imperfections of Bohr's earlier theory had been the consequence of assuming too simple a model for the atom. For Bohr had not only assumed that the atom consisted of particles moving in space and time, but also that the particles inside atoms were of the same kind as the electrons outside atoms.

Now the electron can never be seen directly. The nearest approach to this is in the Wilson cloud-chamber, where we may see the condensation trail which an electron leaves behind it as it pushes its way through the molecules of gas, much as we see the condensation trail left by an aeroplane high up in the sky when we cannot see the aeroplane itself. There is much more evidence of a similar kind, but all of it refers only to electrons outside atoms; the electron inside the atom remains unobserved and unobservable, and there is no solid justification for supposing that it resembles the electrons we see (or so nearly see) outside. We may watch the sparks fly as the blacksmith hammers a piece of iron into a horseshoe, but we must not infer that the piece of iron is an accumulation of sparks, each having the properties of those we see flying through the air.

Bohr's investigation had typified what had become a standard procedure in problems of theoretical physics. The first step was to discover the mathematical laws governing certain groups of phenomena; the second was to devise hypothetical models or pictures to interpret these laws in terms of motion or mechanism; the third was to examine in what way these models would behave in other respects, and this would lead to the prediction of other phenomena—predictions which might or might not be confirmed when put to the test of experiment. For instance, Newton had explained the phenomena of gravitation in terms of a force of gravitation; a later age had seen the luminiferous ether introduced to explain the propagation of light and, subsequently, the general phenomena of electricity and magnetism; finally Bohr had introduced electronic jumps in an attempt to explain atomic spectra. In each case the models had fulfilled their primary purpose, but had failed to predict further phenomena with accuracy.

Heisenberg now approached the problem from a new philosophical angle. He discarded all models, pictures and parables, and made a clear distinction between the sure knowledge we gain from observation of nature and the conjectural knowledge we introduce when we use models, pictures and parables. Sure knowledge, as we have already seen, can only be numerical, so that Heisenberg's results were inevitably mathematical in form, and could not disclose anything about the true nature of physical processes or entities.

As Heisenberg was concerned primarily with the problem of atomic spectra, he found his main observational material in a mass of measurements on the frequencies of the light emitted by the atoms of the chemical elements.

A great deal of regularity had already been detected in these numbers. In 1908 Ritz had noticed that they were the differences of a set of even more fundamental frequencies, being of the form $a-b$, $b-c$, $a-c$, etc. where $a$, $b$, $c$, ... were the more fundamental frequencies. These fundamental frequencies were further known to fall into groups, the numbers in any one group being associated with the series of integers 1, 2, 3, 4, .... Bohr had further discovered that the frequencies corresponding to very large integers could be calculated accurately from the classical mechanics; they were simply the number of times that an ordinary electron would complete the circuit of its orbit in one second when it was at a very great distance from the nucleus of the atom to which it belonged. This could only mean that when an electron receded to a great distance from the nucleus of its atom, it not only assumed the properties of an ordinary electron, but also behaved as directed by the classical mechanics. Yet the classical mechanics failed completely for the calculation of frequencies corresponding to small orbits.

A similar situation had occurred in astronomy, where the Newtonian law of gravitation had been found to predict the orbits of the outer planets with great accuracy, but had failed with the orbits of Mercury and Venus. The relativity theory of gravitation had provided the needed modification of Newton's law, and in working out the details of the new theory, Einstein had utilized the fact that the Newtonian law gave the right result at great distances from the sun. Heisenberg, confronted with a similar problem, was able to avail himself of the fact that the classical mechanics gave the right result at great distances from the atomic nucleus. Here, and here alone, Heisenberg's theory made contact with the world of the older physics. For the classical mechanics was based on the conception of particles moving in space, so that through it Heisenberg's theory entered into relation with space, motion and material particles.

Thus in the outer regions of the atom, Heisenberg's theory coincided both with the classical mechanics and with the newer theory of Bohr. In the interior of the atom, Bohr had tried the plan of retaining the particle-electron and modifying the classical mechanics. Heisenberg took the opposite course, his procedure amounting in effect to retaining the classical mechanics, at least in form, and modifying the electron. Actually, the electron dropped out altogether: it had to, because it exists only as a matter of inference and not of direct observation. For the same reason, the new theory contains no mention of atoms, nuclei, protons, or of electricity in any shape or form. The existences of all these are matters of inference, and Heisenberg's purely mathematical theory could no more make contact with them than with the efficiency of a turbine or with the price of wheat.

The main result reached by the new theory was that the classical mechanics can be made to account for the whole range of spectral phenomena, provided entirely new meanings are given to such symbols as the $p$ and $q$ which had hitherto been taken to describe the position and motion of an electron. The things represented by these symbols acquire new properties which make it impossible that they should any longer represent the simple momentum and distance of a moving particle. In fact, they cease to be mere quantities of any kind, each becoming a whole group of quantities.

The most significant of the new properties is that the product $pq$ is no longer the same thing as the product $qp$—in other words, the order in which the two factors are multiplied together is no longer a matter of indifference. The difference between $pq$ and $qp$ is found to be always the same, being Planck's constant $h$ multiplied by a numerical multiplier.

This last relation in combination with the canonical equations, which are taken over complete from the classical mechanics, provide sufficient mathematical relations for the solution of any problem of quantum mechanics, and, so far as is at present known, invariably lead to the correct solution. Here, then, so far as we can at present see, the true description of the pattern of events must lie.

It may be thought that there is one relation more in the quantum than in the classical mechanics, namely that just mentioned, which gives the value of $pq - qp$. But this is not so; $pq - qp$ has one value in the quantum mechanics and a different value, namely zero, in the classical mechanics. The real difference is that the value of $pq - qp$ is mentioned explicitly in the quantum mechanics, but not in the classical mechanics, where $p$ and $q$ are tacitly assumed to be of such a nature that $pq$ must be equal to $qp$.

Even when this is agreed, it may still seem that the quantum mechanics must represent a complete break with the classical mechanics, since $pq - qp$ has entirely different values in the two systems. But again this is not so. Suppose we use the quantum mechanics to solve a problem on the man-sized scale; $p$ and $q$ are now so large that $pq$ is an immense multiple of $h$, and so also of $pq - qp$. But this is only to say that, to a very close approximation, $pq$ may be taken equal to $qp$, and we are brought back to the classical mechanics.

Thus in problems in which $pq$ is a large multiple of $h$, the quantum mechanics necessarily gives the same result as the classical mechanics, while in problems in which $pq$ is not a large multiple of $h$, it provides a genuine extension of the classical mechanics. Heisenberg's quantum mechanics is universally true, and the classical mechanics is merely a special case of it.

When a problem is solved by the classical mechanics, the solution we obtain depicts continuous motion and change; when it is solved by the quantum mechanics, the solution tells us of jumpy motions and changes of the kind we have already met in Bohr's theory of the hydrogen atom—if the solutions of the classical mechanics describe a ball rolling down an inclined plane, those of the quantum theory describe it as bumping down a staircase. The amount of each jump is proportional to $h$, so that in problems in which $pq$ is a large multiple of $h$, each jump is so small compared with the main motion that the succession of jumps becomes indistinguishable from continuous motion. In this way the jumps of the quantum theory merge into the continuous motion of the Newtonian mechanics.

## Pictorial Representations

If, as now appears fairly certain, Heisenberg's system describes the true pattern of events, it is natural to inquire whether any pictorial representation of the system can be obtained.

The simplest course is to try to imagine that $p$ and $q$ still specify the position and momentum of a moving something, this unknown something becoming identical with the familiar electron when it is at great distances from the atomic nucleus, but this is of no real value, since our minds cannot imagine any kind of structure for which $pq$ would be different from $qp$. If we wish to obtain a really helpful representation, our primary problem must be to find some interpretation of $p$ and $q$, such that the order in which $p$ and $q$ are compounded shall not be a matter of indifference. The simplest procedure is to picture $p$ and $q$ as some sort of operators, since the order in which operations are performed is not usually a matter of indifference. Fining a man £100 and then confiscating half his fortune is not the same thing as confiscating half his fortune and then fining him £100. The difference to the victim is £50, and this corresponds to the value of $pq - qp$ on Heisenberg's theory.

At an early stage in the development of the theory, Born and Wiener found very simple operators which satisfied the requirement that $pq - qp$ should be equal to a constant quantity. But before this had occurred, other attempts to improve on Bohr's theory had resulted in yet another form of the quantum theory, the form which is usually described as the wave mechanics. This was of a much more physical nature than the abstract mathematical theory of Heisenberg, and led to a picture of atomic processes which was not altogether unlike that presented by Bohr's earlier theory.

Now the replacement of $p$ and $q$ in Heisenberg's theory by the operators just mentioned was found to lead exactly to the equations which had already been found to express the wave mechanics. The wave mechanics accordingly falls naturally into place as a pictorial representation of the more general quantum mechanics of Heisenberg. In its mathematical implications, it can be shown to be completely equivalent to Heisenberg's quantum mechanics, and has shown itself able, in principle, to solve every problem so far solved by the quantum mechanics. But we must be on our guard against supposing that the two are exactly equivalent; it must always be remembered that Heisenberg's theory consists of a statement of facts in abstract mathematical form, whereas the wave mechanics consists of a pictorial representation of these facts in which the pictorial details may or may not correspond truly to the realities of nature.

Before proceeding to describe this wave mechanics, it will be convenient to mention some experimental results which are of importance for its understanding.

## ELECTRON WAVES

As science dug ever deeper into the structure of matter, molecules, atoms and electrons were discovered in turn. The last of these appears to be final; no one has ever found a fraction either of the electron or of the electronic charge.

A current of electricity, such as carries our telephone messages or rings our electric bells, consists of a stream of electrons all moving in the same direction. Such currents can not only be passed through solids, liquids and gases, but also through empty space. In this last case, it can be arranged that the electrons shall all move in parallel paths and at the same speed; they may then be described as a shower rather than as a current.

If a thin layer of metal is placed in the path of such a shower, some at least of the electrons of the shower must strike the nuclei and electrons of the atoms of the metal. As they will strike at all sorts of angles, we might expect to find that their courses would be deflected much as bagatelle-balls are deflected by the pins of the bagatelle-board, so that they will emerge at the far side of the metal film as a disordered mob of electrons.

The actual course of events is very different. Part of it was discovered, almost by accident, by two American physicists, Davisson and Germer. They were intending to study the law of scattering of electrons at metal surfaces, and were projecting a shower of parallel-moving electrons on to a sheet of nickel, when their apparatus broke. In the process of mending it, they made their nickel surface so hot that it crystallized.

Now crystal surfaces possess very special properties. The atoms of a non-crystalline substance are not arranged in any regular formation, but are thrown together as though at random, like the grains in a pile of sand. But the atoms of a crystalline substance are arranged in perfect regularity, forming a repeating geometrical pattern of squares, triangles and so forth—a property which has been of great value to experimental physics.

The properties of light are often studied by using a piece of apparatus called a diffraction-grating—a metal plate having parallel lines scratched on its surface with the utmost regularity and precision at the rate of 15,000 to 40,000 to the inch. When a beam of light is reflected from such a surface, it is sorted out into its different spectral colours, much as though it had been passed through a spectroscope. The closer the lines are drawn on its surface, the shorter the wave-lengths of the light with which the apparatus can deal, because the grating becomes ineffective if the distance between successive lines is much greater than the wave-lengths of the light. Red light has about 30,000 waves to the inch, violet light about 60,000. It is easy to rule the lines on a grating close enough to deal with such radiation as this.

On the other hand X-radiation has hundreds of millions of waves to the inch, so that a grating could only cope with this if its lines were ruled at only atomic distances apart. It is obviously impossible to rule lines as close as this by mechanical methods, but some experiments by Laue (1912) showed that it is also unnecessary, since quite perfect gratings of this kind already exist in the surface of crystals, in which the atoms are arranged in perfectly regular formation.

Innumerable experiments have shown that the ridges and depressions formed by these regular chains of atoms cause the surface to act as a natural diffraction-grating for radiation having the wave-length of X-rays. This has opened up new fields of scientific investigation. Sir W. H. Bragg and Sir W. L. Bragg, together with an army of other investigators, have studied the arrangement of atoms in solids by noticing how X-radiation is treated when it falls on the solids, while Siegbahn and others, measuring the wave-lengths of the X-rays emitted by atoms of the various chemical elements, have gained valuable information as to the internal structure of these atoms.

We can now understand what happened when Davisson and Germer shot electrons on to the surface of their crystallized nickel. They found that the reflected electrons were not scattered at random, but showed marked preferences for certain directions in space. They saw that this must result from the regular spacing of the atoms in the nickel surface, but unhappily the electrons they were using moved too slowly for their investigation to be carried to its proper conclusion.

Shortly afterwards Prof. G. P. Thomson performed similar experiments, using faster electrons and improved methods. He made thin films, only about 100 atoms in thickness, out of metals which were naturally crystalline; these were strong enough to hang together, and yet so thin as to be almost transparent. Electrons which moved at about 50,000 miles a second were found to penetrate through these films, instead of being reflected back at their surfaces, and could be made to record their positions after penetration on a photographic plate. The imprints were found to show extreme orderliness in their arrangement; they formed a pattern of concentric circles, light and dark circles alternating, round the point at which the shower of electrons would have struck the plate had the film of metal not been in their way. This showed that the film does not throw the electron formation into disorder, but spreads it in a very regular way. The pattern was found to be the same as would have been formed by X-rays of a certain definite wave-length passing through the same film of metal. If the film is replaced by one of some other substance, the pattern is replaced by the pattern which the same X-rays would form if they passed through the new substance.

We may be tempted at this stage to imagine that the regular pattern is simply impressed on the shower of electrons by the regular arrangement of the atoms in the crystal. But this cannot be the whole cause of the scattering; if it were, passing the shower of electrons through two plates of the solid in succession would produce twice as much scattering as passing it through one. Instead of doing this, it merely lessens the intensity of the pattern, thus proving that the pattern must be produced by some property inherent in the electrons, which is brought to light by their passage through the metal film. This is further shown by the fact that the electrons can be reflected from the metal surface, and still show the same sort of regular pattern.

In each case, the pattern is the same as would be produced by X-rays, so that the shower of electrons must have something in common with X-radiation, and so must possess some quality of an undulatory kind. It is of course a mere numerical accident that in all the experiments the electrons behave like one special type of radiation, namely X-radiation; this results from X-radiation being the only type of radiation with a wave-length comparable with interatomic distances.

If the speed of the electron shower is changed, the pattern changes to one that would be produced by X-radiation of a different wave-length—the slower the electrons, the longer the waves of the equivalent radiation. This wave-length is found to be inversely proportional to the speed of the electron shower, the product of wave-length and speed being equal to $h$ (Planck's constant) divided by $m$ (the mass of the electron). The appearance of Planck's constant here clearly suggests that the wave properties of the electron must in some way be connected with the quantum theory; indeed de Broglie had predicted the relation we have just mentioned from pure quantum considerations, before the wave-pattern had ever been observed.

Such are the purely experimental results. In the preceding chapter we saw how radiation, which was once thought to be wholly undulatory, can be pictured as possessing some of the properties of particles—a beam of radiation falling on a material surface may be pictured as a shower of photons, each located at a definite point of space and possessing mass and energy. We now find that a shower of electrons, which was once thought to consist wholly of particles, may be pictured as possessing some of the properties of waves, at least to the extent of having a definite wave-length associated with it.

### WAVE MECHANICS

These waves form the subject-matter of the wave mechanics, and at the same time, as we have seen, provide a pictorial representation of Heisenberg's quantum mechanics. The fact that the mathematical wave-lengths (although never physical waves) show their presence experimentally provides confirmation both of the truth of the quantum mechanics and of the validity of the wave mechanics as a pictorial representation of it.

When we study the properties of these waves further, we find that they are very similar to the waves of the undulatory theory of light. We have already seen that these latter may be described as waves of probability, the intensity of the waves at any point giving a measure of the probability of a photon occurring at the point. Electron waves may be interpreted in a precisely similar way.

To see this, we need only imagine that in the experiments just described, the strength of the shower of electrons is reduced until it consists of one solitary electron. This, being indivisible, must strike the photographic plate at one and only one point. This point must be one which was darkened in the original experiment, otherwise we should have to suppose that one electron could do what millions failed to do. The darker the plate was then made at any point, the more electrons struck the plate here, and so the greater the chance that the single electron shall strike here now. Thus electron waves, just like waves of radiation, may be interpreted as waves of probability, their intensity at any point giving a measure of the probability of an electron being found at the point.

According to Planck's original theory, a photon is a store of energy of amount equal to $h$ times the frequency of its waves. Now an electron is also known to be a store of energy of amount $mc^2$, where $m$ is the mass of the electron and $c$ is the velocity of light. The general principles of the quantum theory suggest that here also the energy will be $h$ times the frequency of the waves, so that the frequency of the electron waves will be $mc^2/h$.

This means that $mc^2/h$ complete waves pass over any assigned point in a second, and as each wave is of length $h/mu$, the total length of waves passing over any assigned point in a second will be $mc^2/h \times h/mu$, or $c^2/u$. Thus the electron waves travel at a speed $c^2/u$.

This result seems at first very surprising. It is a well-established result of physics that nothing material can travel faster than light. Thus $u$, the speed of the material electron, must be less than $c$, the speed of light, so that $c^2/u$, the speed of the electron waves, must be *greater than the speed of light*. This would sufficiently show, if we did not know it already, that these waves transport nothing material with them. Probability is of course not material, being endowed with neither mass nor energy.

Since the electron waves travel faster than light, it looks at first as though they would rapidly run away from their electrons. Yet this would involve an obvious absurdity. For if electrons travel through space at a speed $u$, the regions where we are likely to find them—i.e. the regions defined by the presence of waves—must obviously travel at the same speed $u$. Actually, as we shall now see, these regions actually do travel only at the speed $u$; the proof of this turns on a somewhat technical point in the general theory of wave-motion.

For purposes of mathematical discussion, the simplest wave system is a train of perfectly regular waves extending for an infinite distance in every direction. Each wave is of precisely the same shape and length, its contour being that of a ripple on still water. Out of a combination of such units, we can build up any formation of waves, no matter how complicated. Conversely of course any wave-formation—as, for instance, a storm at sea—can be analysed into a number of these simple units. The storm may be confined within a circle of 100 miles radius, but each unit must be supposed to extend to infinity in every direction. Outside the storm circle, the waves of the various units still exist in a mathematical sense, but destroy one another by interference, a point which is a crest on one set of waves being a trough on another and so on, in such a way that the total elevation of the surface of the water at every point is *nil*, and the sea is calm.

When the original cause of the storm—the whipping-up of waves by the friction of wind on water—has subsided, each wave-unit pursues its natural motion over the sea as though the other wave-units did not exist. When the motion is traced out mathematically, two distinctive features emerge. In the first place, the interference of waves outside the storm circle becomes less complete as the motion progresses, so that the roughness of the sea gradually extends to areas outside the circle. Also the shorter waves are destroyed more rapidly than the longer by the action of dissipative forces, so that finally only the long waves are left, and we have rollers or a long swell prevailing over the whole ocean.

A somewhat different application of the theory is of special interest to our present problems. By combining a number of units having wave-lengths nearly equal to any assigned wave-length $\lambda$, a wave-formation can be built up which will consist entirely of waves having the precise wave-length $\lambda$, and will extend only over a small region of space. As before, the waves outside this small region of space destroy one another by interference. A short sequence of waves of this kind is called a *wave-packet*.

Let us now imagine each of the constituent units of a wave-packet travelling through space in the way appropriate to its wave-length. It is common in nature for waves to travel at a speed depending on their wave-length, and in the present case each train of waves will travel nearly but not quite at the speed appropriate to the wave-length $\lambda$. We might expect that the whole of the wave-packet would travel at approximately the same speed, but mathematical analysis shows that it does not. In front of the wave-packet, the waves are continually destroying one another by interference, while at the back the reverse process is taking place. This results in a slowing down of the speed of the wave-packet as a whole, so that it advances more slowly than the individual waves of which it is constituted. Detailed analysis shows that, although each individual wave travels at a speed $c^2/u$, the packet as a whole travels only at a speed $u$, which is precisely the speed of the electron. Thus the waves as a whole do not run away from the electron.

We saw that radiation cannot suitably be pictured as particles when it is travelling through empty space. There is a corresponding property for electrons; these should not be pictured as waves so long as they are travelling through empty space. The reason is that the quantities $h/mu$ and $c^2/u$ which specify the waves have no meaning until $u$ can be defined, and, as the theory of relativity shows, it is meaningless to define $u$ as the speed at which electrons are travelling through empty space; it can only be defined in connection with some frame of reference, as for instance some material surface on which the electrons are falling or about to fall. Thus we must think of the electron waves as springing into existence when a current of electricity enters into relation with a material surface, just as we think of photons as springing into existence when radiation meets a material surface.

All this shows that the waves cannot have any material or real existence apart from ourselves. They are not constituents of nature, but only of our efforts to understand nature, being only the ingredients of a mental picture that we draw for ourselves in the hope of rendering intelligible the mathematical formulae of the quantum mechanics. The mathematical specification of the waves is unalterably fixed, being the equivalent of the formulae of the quantum mechanics. But the details of the physical picture are not unalterably fixed. If this picture were perfect, it would enable us to comprehend the incomprehensible, so that we cannot expect it to be very perfect; it may well show some want of precision, and may even be adjusted to meet the special circumstances of a particular problem. Often, for instance, it is a convenience to imagine the electron waves as existing in empty space, just as it may occasionally be convenient to imagine photon waves existing in matter.

Nevertheless, all waves tend to spread—like the waves of our storm at sea, or ripples on the surface of a pond. Whether a wave-packet is large or small, it must continually increase in size, and the smaller it is to begin with, the more rapidly it grows. This shows that no wave-packet can permanently represent a single electron; an electron is a permanent structure, while a wave-packet is not. Indeed the wave mechanics has no concern with single electrons. But we may import the concept of the atomicity of electricity into it from experimental physics, and we then notice that if any wave-packet represented an electron at one moment it would have ceased to do so by the next, because the wave-packet would have changed, while the electron had not.

We might perhaps conjecture that different wave-packets represent electrons in different circumstances. If so, let us see whether we can discover the circumstances of the electron corresponding to a few simple types of wave-packet.

Suppose first that the wave-packet is of only infinitesimal length, a mere point in space. Such a wave-packet might seem specially suited to represent an electron under most circumstances. But it is a mathematical fact that a wave-packet of only infinitesimal length cannot have, or even be associated with, any definite wave-length; there is not room, so to speak, for the wave qualities to develop. We have seen that a packet of wave-length $\lambda$ represents an electron moving with a speed $h/m\lambda$, so that, if we cannot form any idea of the value of $\lambda$, we are equally unable to form any idea of the speed of the electron.

If the length of the wave-packet is gradually increased, definite wave qualities gradually emerge. Finally the packet becomes an endless train of waves, each of wave-length equal to the wave-length of the packet. If an electron is represented by such an infinite train of waves, we can of course determine its speed of motion with absolute precision; it is simply $h/m\lambda$, and there is now no uncertainty about the value to be assigned to $\lambda$. But now we are totally unable to say where the electron is. The wave-packet has become an endless and featureless train of waves, and there can be no reason for assigning the electron to any one point of it rather than to another. Thus we see that a short train of waves would fix the position of the electron in space, but would fail to fix its speed of motion, while a long train of waves would tell us the speed of motion, but could not fix the position of the electron in space. No conceivable wave-packet could indicate both the speed of motion and the position of an electron with absolute precision.

This immediately reminds us of a result we obtained in Chapter IV (p. 142). We there saw that our experimental explorations of nature do not admit of absolute precision, owing to the fact that nothing less than a complete photon can be received from the outer world. Regarding the electron as a moving particle, we saw that no experiment could fix both its speed of motion and its position in space with complete accuracy. If quanta of low frequency are used in an experiment, the determination of the electron's position will necessarily be very uncertain, while if quanta of high frequency are used, the uncertainty is merely transferred to the determination of the electron's momentum, since the energetic photon gives the electron a big kick in leaving it. No possible experimental arrangement can make these two uncertainties both vanish simultaneously, so that the product of the two can never be zero. A detailed study by Heisenberg has shown that the product can never be less than

Planck's constant $h$. This is known as Heisenberg's principle of uncertainty (or indeterminacy).

We have just seen that the wave-packet of an electron shows a precisely similar lack of precision. Again a detailed mathematical discussion shows that whatever kind of wave-packet we select to represent an electron, the product of the two uncertainties of position and momentum can never be less than $h$, which is precisely what Heisenberg found about the experimental investigation.

When an electron is depicted as a particle in space, it has an exact speed of motion and also an exact position in space, both of which can be specified by numerical quantities; the trouble revealed by the uncertainty principle is not that these quantities do not exist, but that we have no practical means of measuring them; they can exist in the electron, but not in our knowledge of the electron. But when we depict the electron as a wave-packet, these quantities do not even exist in the wave-packet.

As Bohr was the first to point out, this gives us the clue to the whole situation and lets the secret out—different kinds of wave-packet must not be supposed to represent different kinds of electrons, or electrons in different states, or electrons under different conditions, but the different kinds of knowledge we can have about electrons. Indeed, just as the waves of the undulatory theory of light were found to represent our knowledge about photons (p. 136), so the waves of the wave mechanics are now seen to represent our knowledge about electrons. Both sets of waves are mental constructs of our own; both are propagated in conceptual spaces.

There is complete parallelism except in one respect. The waves of the undulatory theory need a space of only three dimensions for their representation, so that we may conveniently and legitimately represent them in ordinary physical space. The waves of a single electron can also be represented in a space of three dimensions, but the waves of two electrons need a space of six dimensions, three for each electron, while the waves of a million electrons need a space of three million dimensions. Thus the wave-picture of even the simplest group of electrons, or of other particles, cannot be drawn in ordinary space.

The wave-picture just described is due to de Broglie, Schrödinger,

Bohr and Heisenberg. It is subjective in the sense that it may depend on experiments that we have recently performed on electrons, but it is also objective in the sense that it shows a capacity at least equal to that of the particle-picture to interpret objective reality, giving correct solutions to many problems in which the particle-picture fails. Indeed, in its mathematical formulation the wave-picture is exactly equivalent to Heisenberg's scheme which, from the mode of its derivation, is necessarily true to reality.

It should, however, be added at once that those cases in which the wave-picture meets with more success than the particle-picture are not those in which it represents the knowledge of any particular individual. The majority have to do with the spectra of atoms, and so are concerned with the motion of electrons round nuclei, and not in free space. The wave-packet still represents our knowledge of the electron, but it is now knowledge as to the possible or probable positions of the electron inside the atom, knowledge which is independent of any particular observation or observer. The wave-packet of a free electron represents private knowledge, individual to a particular observer who has recently made an observation on the electron, but the wave-packet of an electron inside an atom represents public knowledge, accessible to all without an experiment. An observer could of course find out more about the electron inside the atom by a new *ad hoc* experiment—as for instance by bombarding the atom with α-particles and noting the Wilson-chamber condensation trail of the electron as it was shot out of the atom—but he would destroy the atom in so doing. The wave-packet of the electron would be concentrated into a smaller region, and would become the wave-packet of a free electron starting off on a new journey.

Thus there is a standard wave-packet for an electron inside an atom, or rather there are several distinct standard packets—one for each state of steady motion which can take place inside the atom—but there is no standard wave-packet for an electron travelling freely through space. This reminds us of what we found in discussing the pictures of an electron provided by the classical mechanics. We found a bullet-picture which corresponded to our present wave-picture, and a 'tentacle'-picture which corresponded to our present wave-picture, but there was no standard tentacle-picture suited for all circumstances. The appropriate picture depended on the motion, not only of the electron but of other bodies as well.

If the waves of a free electron or photon represent human knowledge, what happens to the waves when there is no human knowledge to represent? For we must suppose that electrons were in existence while there was still no human consciousness to observe them, and that there are free electrons in Sirius where there are no physicists to observe them.

The simple but surprising answer would seem to be that when there is no human knowledge there are no waves; we must always remember that the waves are not a part of nature, but of our efforts to understand nature. Whether we are thinking about electrons or not, and whether we are experimenting with them or not, their motion is determined by the equations of the Heisenberg dynamics. When an electron joins an atom or is knocked out of an atom, its motion undergoes just the same changes whether we are presiding over the experiment or not; if a photon is emitted, it makes no difference to the electron whether this photon ends up in a human eye or elsewhere.

Similar remarks may be made about the waves of the undulatory theory of light, and about the electric and magnetic forces of which we have hitherto imagined them to be constituted. Energy may be transferred from place to place, but the waves and the electric and magnetic forces are not part of the mechanism of transfer; they are part simply of our efforts to understand this mechanism and picture it to ourselves. Before man appeared on the scene, there were neither waves nor electric nor magnetic forces; these were not made by God, but by Huyghens, Fresnel, Faraday and Maxwell.

## DIRAC'S QUANTUM MECHANICS

The third form of quantum mechanics, that of Dirac, must be dismissed very briefly, not because it is in any way unimportant, but because it is so intensely mathematical in form as to lie entirely beyond the scope of the present book. Dirac's ambition was to put the whole of quantum mechanics in a perfectly consistent form, deducing all its conclusions from a few simple assumptions much as Euclid deduced the whole of geometry from a few simple axioms.

Dirac remarks that the classical mechanics had tried to explain physical phenomena in terms of particles and radiation moving in space and time; it made a few simple assumptions about the factors governing the bodies which figure in the phenomena, and then tried to account for their behaviour in terms of these assumptions. In brief, it tried to explain the phenomena without going beyond the phenomena, as though the world of phenomena formed a closed whole. This attempt failed, and it became clear that nature works on a different plan. Exhaustive studies by many investigators have shown that the fundamental laws of nature do not control the phenomena directly. We must picture them as operating in a substratum of which we can form no mental picture unless we are willing to introduce a number of irrelevant and therefore unjustifiable suppositions.

Events in this substratum are accompanied by events in the world of phenomena which we represent in space and time, but the substratum and the phenomenal world together do not form a complete world in itself which we can observe objectively without disturbing it. The complete closed world consists of three parts—substratum, phenomenal world, and observer. By our experiments we drag up activities from the substratum into the phenomenal world of space and time. But there is no clear line of demarcation between subject and object, and by performing observations on the world we alter it, much as a fisherman, dragging up a fish from the depths of the sea, disturbs the waters and also damages the fish.

Dirac introduces operators of an abstract mathematical kind, to represent the effect of dragging an activity up to the surface—i.e. of observing it. He finds it necessary to assume that the series of observable types of activity $a$, $b$, $c$,... is more restricted than the corresponding series of types in the substratum. The latter series consists of certain *pure* types $A$, $B$, $C$,... which appear as $a$, $b$, $c$,... in the world of phenomena, and also of certain *composite* types, which we may denote by $AB$, $BC$, $AC$,... and have no

direct counterparts in the phenomenal world. *AB* may give rise
to *a* or to *b*, but never to both, and there is an assignable probability
as to whether *a* or *b* will appear. Thus the substratum of reality is
in some way richer and more varied than the world of phenomena.

After elaborate mathematical discussion, Dirac reaches a formal
theory of a very complete kind. The matrix mechanics of Heisenberg and the wave mechanics of de Broglie and Schrödinger are
then shown to be included in the theory as special cases.

It will be seen from this that the pattern of events implied in
Dirac's theory is necessarily the same as the pattern implied in the
theories of Heisenberg, and of de Broglie and Schrödinger, and
so agrees entirely with that observed in nature. It is an essential
feature of Dirac's theory that events in the phenomenal world are
not uniquely associated with events in the substratum; different
events in the substratum may result in phenomena which are precisely similar, at least to our observation. Thus the same phenomenon in the space-time world may be associated with a number
of different states in the substratum, and so may be followed by
different events. Because of this, experiments which are similar so
far as our observation goes need not, and usually will not, lead to
identical results. Thus the uniformity of nature is jettisoned at the
outset in so far as the phenomena are concerned, and causality
disappears from the world we see.

It does not entirely disappear from the world which is hidden
from our view. The mathematical equations of both forms of the
new quantum theory—the wave mechanics and the matrix mechanics—are completely deterministic in form. So far as these
equations go, the future of the world appears to be a mere unrolling
so that the future follows uniquely and inexorably from the past.
But this unrolling is not, as we have already seen, of the course
of events, but of our knowledge of events. Causality disappears
from the events themselves to reappear in our knowledge of events.
But, since we can never pass behind our knowledge of events to
the events themselves, we can never know whether causality
governs the events or not. Indeed the considerations mentioned
on p. 144 suggest that even to discuss the question is meaningless.

# CHAPTER VII

## SOME PROBLEMS OF PHILOSOPHY

We have now concluded our summary of the findings of modern physics, and may turn to consider how these findings affect the practical problems of philosophy and of everyday life. But let us first recapitulate the conclusions we have reached in our scientific discussion.

### Recapitulation

Because we are human beings and not mere animals, we try to discover as much as we can about the world in which our lives are cast. We have seen that there is only one method of gaining such knowledge—the method of science, which consists in a direct questioning of nature by observation and experiment.

The first thing we learn from such questioning is that the world is rational; its happenings are not determined by caprice but by law. There exists what we have called a 'pattern of events', and the primary aim of physical science is the discovery of this pattern. This, as we have seen, will be capable of description only in mathematical terms.

The new quantum theory explained in the preceding chapter has provided a mathematical description of the pattern of events which is believed to be complete and perfect. For it enables us— in principle at least—to predict every possible phenomenon of physics, and not one of its predictions has so far proved to be wrong. In a sense, then, we might say that theoretical physics has achieved the main purpose of its being, and that nothing remains but to work out the details.

But we not only wish to predict phenomena, but also to understand them. Thus it is not surprising that philosophy and science have alike found this mathematical description unsatisfying, and have tried to attach concrete meanings to the mathematical symbols involved—to replace unintelligible universals by intelligible particulars. We may argue that if there is a pattern, there must be some sort of loom for ever weaving it; we want to know what this loom is, how it works, and why it works thus rather than otherwise.

The physicists of the last century thought that one of the primary concerns of science should be to devise models or draw pictures to illustrate the workings of this loom. It was supposed that a model which reproduced all the phenomena of a science, and so made it possible to predict them all, must in some way correspond to the reality underlying the phenomena. But obviously this cannot be so. After one perfect model had been found, a second of equal perfection might appear, and as both models could not correspond to reality, we should have at least one perfect model which did not correspond to reality. Thus we could never be sure that any model corresponded to reality. In brief, we can never have certain knowledge as to the nature of reality.

We know now that there is no danger of even one perfect model appearing—at least of a kind which is intelligible to our minds. For a model or picture will only be intelligible to us if it is made up of ideas which are already in our minds. Of such ideas some, as for instance the ideas of abstract mathematics, have no special relation to our particular world; all those which have must, as we have seen, have entered our minds through the gateways of the senses. These are restricted by our having only five senses of which only two are at all important for our present purpose.

A detailed investigation of the sources of our ideas has shown that there is only one type of model or picture which could be intelligible to our restricted minds, namely one in mechanical terms. Yet a review of recent physics has shown that all attempts at mechanical models or pictures have failed and must fail. For a mechanical model or picture must represent things as happening in space and time, while it has recently become clear that the ultimate processes of nature neither occur in, nor admit of representation in, space and time. Thus an understanding of the ultimate processes of nature is for ever beyond our reach; we shall never be able—even in imagination—to open the case of our watch and see how the wheels go round. The true object of scientific study can never be the realities of nature, but only our own observations on nature.

### The Particle-picture and the Wave-picture

Although there can be no complete picture of the workings of nature which will be intelligible to our minds, yet we can still draw pictures to represent partial aspects of the truth in an intelligible way. The new physics places two such partial pictures before us—one in terms of particles, and one in terms of waves. Neither of these can of course tell the whole truth.

In the same way, an atlas may contain two maps of North America drawn on different projections: neither of them will represent the whole truth, but each will faithfully represent some aspect of it. An equal-area projection, for instance, represents the relative areas of any two regions accurately, but their shapes wrongly, while a Mercator projection represents the shapes rightly, but the areas wrongly. So long as we can only draw our maps on flat pieces of paper, such imperfections are inevitable; they are the price we pay for limiting our maps to the kind that can be bound up in an atlas.

The pictures we draw of nature show similar limitations; these are the price we pay for limiting our pictures of nature to the kinds that can be understood by our minds. As we cannot draw one perfect picture, we make two imperfect pictures and turn to one or the other according as we want one property or another to be accurately delineated. Our observations tell us which is the right picture to use for each particular purpose—for instance, we know we must use the particle-picture for the photo-electric effect, the wave-picture for illumination effects, and so on.

Yet some properties of nature are so far-reaching and general that neither picture can depict them properly of itself. In such cases we must appeal to both pictures, and these sometimes give us different and inconsistent information. Where, then, shall we find the truth?

For instance, is nature governed by causal laws or not? The particle-picture answers: No, the motions of my particles can only be compared to the random jumps of kangaroos, with no causal

laws controlling the jumps. But the wave-picture says: Yes, at
every instant my waves follow uniquely, and so inevitably, from
those of the preceding instant.

Or again, is reality ultimately atomic or is it not? The particle-
picture tells us of a material world in which matter, electricity and
radiation occur only in indivisible units; the wave-picture merely
tells us that it knows of none of these things.

The two pictures seem to tell different stories, but we must
remember that they are not equally trustworthy. The particle-
picture embodies the findings of the old quantum theory which we
discussed in Chapter v. This proved to be both inaccurate and incomplete, so that the new quantum theory was brought into being
to remedy its deficiencies—which it has successfully done. The
wave-picture is not only a pictorial representation of the new
quantum theory, but also, as regards the mathematical facts involved, is its exact equivalent. Thus the predictions of the wave-
picture cannot be other than true, whereas those of the particle-
picture may or may not be true. When there is a conflict, the
evidence of the wave-picture must be accepted, while we may be
sure that the conflict results from some imperfection of the particle-
picture. In the instances just given, it is not difficult to trace out
a possible origin for the conflict.

The mathematical laws of the quantum theory show that radiant
energy is transferred by complete quanta. But in depicting a
beam of light as a hail of bullet-like photons, the particle-picture
is clearly going further than the facts warrant. A man's balance at
the bank always changes by an integral number of pence, but this
does not justify him in picturing its changes as caused by a flight
of bronze pennies. If he does, his child may ask him what decides
which particular pennies shall be sent to pay the rent. The father
may reply: Mere chance—a foolish answer but no more foolish
than the question. In the same way, if we make the initial mistake
of depicting radiation as identifiable photons, we shall have to call
on mere chance to get us out of our difficulties—and here is the
origin of the indeterminacy of the particle-picture.

For instance, when a beam of light falls on a half-silvered mirror
(p. 137), the particle-picture shows half the photons being turned

back by the silvering of the mirror, while the other half pass on their way undisturbed. We ask at once: What singles out the lucky photons? It is a question which had confronted Newton's corpuscular theory of light, and he had answered it by a vague wave of the hand towards Fortune's wheel—his corpuscles, he had said, were 'subject to alternate fits of easy transmission and easy reflection'. In the same way, if we depict radiation as identifiable photons, we can find nothing but the finger of Fate to separate the sheep from the goats. But the finger of Fate, like the sheep and the goats, is mere pictorial detail. As soon as we turn to the more trustworthy wave-picture, all this pictorial drapery drops out of the picture, and we find a complete determinism. Yet this determinism, as we have seen, does not control events, but our knowledge of events. The wave-picture does not show the future following inexorably from the present, but the imperfections of our future knowledge following inexorably from the imperfections of our present knowledge.

What is true of radiation is true also of electricity. We know that electricity is always transferred from place to place by complete electron-units, but this does not justify us in replacing a current of electricity by a shower of identifiable particles. Indeed, the quantum theory definitely tells us that we must not do so. When two balls $A$, $B$ collide on a billiard-table, $A$ may go to the right and $B$ to the left. When two electrons $A$, $B$ collide, we might also expect to be able to say that $A$ would go to the right and $B$ to the left; actually we cannot, because we have no right to identify the two electrons which went into the collision with the two which come out; we must rather think of the two electrons $A$ and $B$ which entered into collision as combining into a drop of electric fluid, which then breaks up again to form two new electrons $C$ and $D$. If we ask which way $A$ will go after collision, the true answer is that $A$ no longer exists. The superficial answer is that it is an even chance whether $A$ goes to the right or to the left, for it is a toss-up whether we identify $A$ with $C$ or $D$. But the toss-up is not in nature; it is in our minds.

We see, then, that the particle-picture goes wrong in attributing indeterminism to nature; it is not a property of nature, but of our way of looking at nature. The particle-picture further goes wrong in attributing atomicity to the ingredients of the material world, whether matter or radiation; the atomicity does not reside in these ingredients but in the events which affect them. To return to our former analogy, all payments into and out of a bank account are by complete mathematical pence, but they do not consist of bronze pennies flying hither and thither. But we can now carry this train of ideas a little further; we know matter only through the energy or particles it emits, but this provides no warrant for assuming that matter itself consists of atoms either of substance or of energy— this would be like assuming that our balance at the bank must consist of a pile of bronze pennies.

## New Philosophical Principles

We have seen that efforts to discover the true nature of reality are necessarily doomed to failure, so that if we are to progress further it must be by taking some other objective and utilizing some new philosophical principles of which we have not so far made use. Two such suggest themselves. The first is the principle of what Leibniz described as *probable reasoning*; we give up the quest for certain knowledge, and concentrate on that one of the various alternatives before us which seems to be most probably true. But how are we to decide which of the alternatives is most likely to be true? This question has been much discussed of late, particularly by H. Jeffreys. For our purpose it is sufficient to rely on what may be described as the *simplicity postulate*; this asserts that of two alternatives, the simpler is likely to be the nearer to the truth.

Let us try to illustrate these new principles by considering a simple, although very artificial, analogy.

Let us imagine that in the centre of Europe there lives a peasant who has never seen or heard of the sea, and cannot even read about it, but is in possession of a super-perfect radio-set which can pick up messages from every ship in the world. Suppose further that every ship is continually sending out its position in a standard form, such as

'Queen Mary', $+41°\ 10'$, $-72°\ 26'$,

this meaning that, at the moment of speaking, the ship 'Queen Mary' is in latitude 41° 10′ north and longitude 72° 26′ west.

At first he may merely amuse himself by listening to the various messages, but after a time he may take to recording them and, if he is of an inquiring turn of mind, he may try to discover some method or order in them. He will soon notice that all latitudes lie between +90° and −90°, and all longitudes between +180° and −180°. If he then tries plotting out these numbers on squared paper, he will find that successive positions of any one ship form a continuous chain, and may begin to construct a mental picture for himself by thinking of the senders of the messages as moving objects. He will then find that each supposed object moves at an approximately uniform rate on his chart, although this law is not exact or universal. A ship may move from longitude +170° to +174° in one day, and on to +178° the next, but the third day may take it to −178°, an apparent journey of 356°. Further, a ship may move at a regular 4° a day when its latitude is near to 0°, but this daily motion will increase as the latitude increases, and may shoot up almost beyond limits if ever the latitude approaches to 90°.

If, notwithstanding their peculiar nature, our listener succeeds in formulating exact laws, he will then be able to predict the motions of the ships. Or, to be more precise, he will be able, without assuming that he is dealing with either motions or ships, to predict what he will hear when he turns on his radio. He can predict the result of every experiment he can perform, since the only experiment within his power is to turn a knob and listen.

Those who are content with a positivist conception of the aims of science will feel that he is in an entirely satisfactory position; he has discovered the pattern of events, and so can predict accurately; what more can he want? A mental picture would be an added luxury, but also a useless luxury. For if the picture did not bear any resemblance at all to the reality it would be valueless, and if it did it would be unintelligible, since we are supposing that our listener cannot imagine either sea or ships.

## Probable Reasoning

At this point, let us notice that the supposition that the signals came from moving objects was hypothetical in the sense that nothing in the observations compelled it—from the nature of the case the observer is debarred from knowing whether the signals come from moving objects or not. It expresses a possibility and not certain knowledge, and can never be proved true. In real science also a hypothesis can never be proved true. If it is negatived by future observations we shall know it is wrong, but if future observations confirm it we shall never be able to say it is right, since it will always be at the mercy of still further observations. A science which confines itself to correlating the phenomena can never learn anything about the reality underlying the phenomena, while a science which goes further than this, and introduces hypotheses about reality, can never acquire certain knowledge of a positive kind about reality; in whatever way we proceed, this is for ever denied us.

Certain knowledge is, however, equally beyond our reach in most departments of life. Oftener than not, we cannot wait for certain knowledge, but order our affairs in the light of probabilities. There is no reason why we should not do the same in our efforts to understand the universe, provided we always bear in mind that we are discussing probabilities and not certainties.

The philosopher does it as much as the rest of us. I am conscious only of my own thoughts and sensations, so that, for aught I know to the contrary, I may be the only conscious being in the universe. If I choose on these grounds to become a solipsist—i.e. one who supposes that he is the only conscious being in the whole universe— nothing can definitely prove me wrong. But my sensations inform me of other objects that look like my body, and seem to experience sensations and thoughts like my own. I assume, although only on grounds of probable reasoning, that these other objects are beings essentially similar to myself. If we refused to admit probability considerations, we ought all to be solipsists; with things as they are, any genuine solipsists there may be are kept safely shut up.

The physicist also relies on probability considerations every day of his life. He measures the wave-lengths of spectral lines in the light emitted by Sirius, and finds they are identical with those in the light emitted by hydrogen at a temperature of 10,000° C. He concludes without more ado that there are atoms of hydrogen at 10,000° in Sirius. There is no proof of this and never can be, for we shall never be able to go to Sirius to find out. But the probabilities against the agreement being a mere coincidence are so overwhelming that the physicist feels justified in disregarding this possibility, and announces that this part of the light of Sirius comes from hydrogen at a temperature of 10,000°.

In these two instances, the philosopher and physicist are both guided by probable reasoning rather than by certain deductions. If our radio listener allows himself to be guided by similar considerations, he may decide provisionally that his signals come from moving objects. This idea may lead him to think of pasting together his $+180°$ and $-180°$ lines, thus transforming his plane diagram into a cylinder. This simplifies the situation enormously, for it now seems the most natural thing in the world that a sequence of readings equidistant in time should read 170°, 174°, 178°, $-178°$, etc. But he is still faced with the peculiarity that his moving objects traverse more degrees of longitude per day in high latitudes than in low. With a little ingenuity, he may further think of crumpling in the two ends of his cylinder, and so making the degrees of longitude smaller in higher latitudes. If he finally tries the experiment of replacing his cylinder by a sphere, he will find that his laws assume an exceedingly simple form from which all oddity has disappeared. Each ship takes the shortest course from point to point, and performs its journey at a uniform speed.

Even the original laws were true laws, since they enabled the listener to predict accurately. But they were not simple, because their discoverer had expressed them against a bad background. As soon as he changed from one background to another—from a rectangular projection to a spherical surface—the laws changed from being strange but true to being simple and true. Precisely for this reason, most men will consider that the second set of laws was preferable. Without assigning any special attributes to the Designer of the universe, we probably feel that the simpler laws are likely to be in some way closer to that reality which we can never understand, than complicated and odd laws—in brief, that artificiality comes from man, and not from nature. In the example just considered, it is certainly more true to say that the earth's surface is spherical than to picture it as plane.

And in the real problems of science also, it is true, as Einstein has remarked, that 'In every important advance the physicist finds that the fundamental laws are simplified more and more as experimental research advances. He is astonished to notice how sublime order emerges from what appeared to be chaos. And this cannot be traced back to the workings of his own mind but is due to a quality that is inherent in the world of perception.'

This not only shows that our minds are in some way in harmony with the workings of nature—a harmony which Einstein compares with the pre-established harmony of Leibniz (p. 27)—but also that our investigations of nature are proceeding on the right lines; it further shows that the simplicity which is inherent in nature is of the kind which *our minds* adjudge to be simple. Indeed any other kind of simplicity would probably escape our notice.

### The Simplicity Postulate

This suggests the introduction of a further principle, if not into the technique of scientific investigation at least into the practice of philosophical discussion—the principle of simplicity. When two hypotheses are possible, we provisionally choose that which our minds adjudge to be the simpler, on the supposition that this is the more likely to lead in the direction of the truth. It includes as a special case the principle of Occam's razor—*entia non multiplicanda praeter necessitatem.*

There can of course be no absolute criterion as to which of two hypotheses is the simpler; in the last resort this must be a matter of private judgment. In the fictitious example we have just been discussing there could be no room for doubt, but in actual scientific practice there have been cases in which two investigators have differed as to which of two hypotheses was the simpler, as for example with the one-fluid and two-fluid theories of electricity.

The history of science provides many instances of situations such as we have been discussing. To begin with the most obvious, Ptolemy and his Arabian successors built up the famous system of cycles and epicycles which enabled them to predict the future positions of the planets with almost perfect precision. At first, the sun, moon and stars were supposed to revolve round the fixed central earth, while the planets revolved about other centres which themselves revolved about the earth. It was soon found that this did not quite fit the facts, and the orbits had to be changed to slightly eccentric circles—neither the earth nor the moving centres were any longer at the exact centres of the circles which were described around them. Finally, as the planetary motions came to be known to a still higher degree of accuracy, epicycle was piled on epicycle until the system became exceedingly complex.

Many, indeed, felt that it was too complex to correspond to the ultimate facts. In the thirteenth century, Alphonso X of Castile is reported to have said that if the heavens were really like that, 'I could have given the Deity good advice, had He consulted me at their creation.' At a later date Copernicus also thought the Ptolemaic system too complex to be true and, after years of thought and labour, showed that the planetary motions could be described much more simply if the background of the motions were changed: Ptolemy had assumed a fixed earth; Copernicus substituted a fixed sun. We know now that neither the earth nor the sun is in any true sense at rest, but we also know why it introduces fewer complications to suppose the sun to be at rest rather than the earth—why it is, in a sense, nearer to the truth to say that the earth moves round the sun than to say that the sun moves round the earth.

We may notice in passing that Copernicus did not claim any absolute truth for his hypotheses, saying that they need not be true or even plausible; it was enough for them to reconcile the observations with the calculations—'neque enim necesse est, eas hypotheses esse veras, imo ne verisimiles quidem, sed sufficit hoc unum, si calculum observationibus congruentem exhibeant'. This reads like a foreshadowing of positivist doctrines, but it may have been only an attempt to propitiate ecclesiastical and religious readers who might have taken fright at the implications of the new hypotheses.

Copernicus had still to retain a few minor epicycles to make his system agree with the facts of observation. This, as we now know, was the inevitable consequence of his assumption that the planetary orbits were circular: neither he nor anyone else had so far dared to challenge Aristotle's dictum that the planets must necessarily move in circular orbits, because the circle was the only perfect curve. As soon as Kepler substituted ellipses for the Copernican circles, epicycles were seen to be unnecessary, and the theory of planetary motions assumed an exceedingly simple form—the form it was to retain for more than three centuries, until an even greater simplicity was imparted to it by the relativity theory of Einstein, to which we shall come in a moment.

The restricted (or physical) theory of relativity provides a second illustration of the same thing. The Newtonian mechanics, with its background of absolute space and time, had explained the motion of objects well enough so long as their speeds of motion were not comparable with that of light. But, as experiment ultimately showed, it could only explain the motion of rapidly moving objects at the price of introducing extreme complications. Objects in rapid motion had to contract and assume new shapes, while no one could ever quite say what happened to objects in rapid rotation. The theory of relativity introduced a tremendous simplification into the whole subject when it discarded Newton's absolute space and time as a background, and substituted the new space-time unity, as explained on p. 63.

The generalized (or gravitational) theory of relativity provides an even more striking instance of the same thing. The Newtonian theory of gravitation, which required the planets to move round the sun in elliptical òrbits, gave an excellent account of the movements of the outer planets, but failed with the inner. Attempts were made to remedy this by slightly altering the Newtonian law of gravitation, by supposing the sun to be surrounded by clouds of gas or dust which impeded the free motion of the inner planets, and in a variety of other ways. The relativity theory of gravitation then cleared up the whole situation at one stroke by rejecting Newton's force of gravitation altogether, and impressing a curvature on the space-time unity in which the motions of the planets were depicted. Once again the change was from an unsuitable to a suitable background. The whole motion of planets and other bodies, as well as of rays of light, could now be described by the simple statement that they all described geodesics—i.e. took the shortest possible course from point to point—in the new curved space-time unity.

The simplification which this change introduced was not only tremendous in itself, but was in line with a number of earlier simplifications, all based on the idea of a length of path or some similar quantity assuming the smallest value which was possible for it.

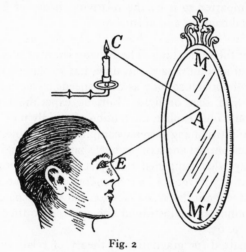

Fig. 2

The principle made its first appearance in optics. If a candle is burning at $C$, and my eye at $E$ looks at a mirror $MM'$, I shall seem to see the candle at some point $A$ in the mirror. This shows that rays of light are travelling along the path $CAE$ from the candle to my eye, and along no others; for if they travelled along any other path $CBE$ as well, I should seem to see candles at both $A$ and $B$, which I do not. Hero of Alexandria set himself the problem of finding what it was that specially distinguished the path $CAE$ which the light actually took from every other possible path such as *CBE* which it might have taken, but did not. He found that *CAE* was the shortest path from *C* to *E* which touched the mirror on its way. Even though the light is reflected from hundreds of mirrors, the path is still determined by the same principle; it is the shortest path that can be found, subject to the condition of its touching all the mirrors in turn. Alternatively the path may be described as the quickest from *C* to *E*; the light chooses its path on the principle of wasting as little time as possible on the way.

Fermat (1601–1665) showed that this latter principle still determines the path when the light travels through water, glass, or other refracting substances of any kind whatever. Thus it is true under all circumstances that light always travels by the quickest route; this provides another instance of the tremendous simplifications to which Einstein refers (p. 183).

Maupertuis (1698–1759) subsequently conjectured that the motions of tangible objects must conform to some similar principle, arguing that Divine perfection would be opposed to any expenditure of energy by moving bodies, beyond the absolute minimum necessary to get from one place to another. In time such a principle was found to govern the motion of all bodies of tangible size—the principle of 'Least Action'. This principle includes the Newtonian mechanics and the classical mechanics as special cases, so that it covers not only mechanical activities but those of electricity and magnetism as well. It can best be understood through a simple analogy.

When I hire a taxicab, the taximeter piles up the charges against me at a rate which depends both on where I am, and on how fast I am travelling. I have to pay one sum per five minutes when I am at rest in a city, some other sum per five minutes when I travel at 15 miles an hour in the city, twice as much when I travel at 30 miles an hour in the city, and so on, and on an entirely different tariff when I am outside the city limits. Now let us imagine a taximeter attached to every moving object in the universe, piling up charges at a rate which depends on both the speed of motion and the position of the object. Let all the objects move for some specified time, such as an hour, and at the end of the motion let all the charges shown on the various taximeters be added up. The principle of Least

Action tells us that the actual objects in nature will have chosen their paths so as to make the total charge shown by all the taximeters a minimum—Nature, setting her face against unnecessary expenditure on taxicabs, always chooses the cheapest route.

Suppose, for instance, that a single particle has to be transferred, within a specified time, from one point $A$ to another point $B$, through a region in which conditions are absolutely uniform, so that the taxicab tariff is of course uniform also. The cheapest way of making the journey will be to travel in a perfectly straight line at a perfectly uniform speed, which is what Newton's law of motion tells the particle to do. Or again, suppose that a planet has to be transported from its present position to the corresponding position at the other side of the sun. The shortest route would be straight through the centre of the sun, but, as the tariff in intense gravitational fields is exorbitant, the charges by this route would be prohibitive. We find we can avoid these excessive charges by taking a curved path round the sun, even though this lengthens the journey somewhat. If part of the route still goes near to the sun, it is cheapest to perform this part of the journey at high speed, so as to spend as little time as possible in the region of exorbitant tariffs. Exact mathematical analysis is needed to find exactly what combination of path and speed reduces the total charge to an absolute minimum; it tells us that the path must be an ellipse having the sun in one of its foci. This is precisely the path demanded by the Newtonian mechanics, but we notice that it is no longer mapped out by the action of 'forces' of the Newtonian kind.

Logically, and to some extent chronologically also, the principle of Least Action forms a direct successor to the principle of Least Time of Hero and Fermat. The principle of Least Distance, or geodesics, in the curved space-time of relativity is clearly in the same line of succession. It introduces a great simplicity by changing to the new background of a curved space—like the change of background of our radio listener when he changed from a rectangular projection to a curved spherical surface. Like the principles of Least Time and of Least Action, this principle of Least Distance shows an extreme simplicity which suggests that we are keeping in close touch with the true significance of natural processes.

The old quantum theory did not show any such simplicity. We need not concern ourselves with it any further since it has now become clear that it was only an unsatisfactory hybrid between the classical mechanics and the new quantum theory, being, in fact, a last desperate effort to represent nature against a background of time and space.

In the new quantum theory the same simplicity reappears in full strength and almost in the same form. So far as its formal mathematical description goes, the theory is a genuine extension of the old Newtonian mechanics, so much so that the same mathematical equations will serve for the description of both, namely the canonical equations of which we spoke on p. 113, these in turn being an expression of the principle of Least Action.

But the pictorial representations that must be given to these equations differ widely in the two cases. The classical mechanics came into existence as an effort to describe the continuous motions of objects under pushes and pulls; it is in this way that it is usually interpreted. But the new quantum mechanics must be interpreted rather as a description of steady states in which either there is no motion or else the state of motion does not change. Now and then, as we have seen, a jump occurs from one of these steady states to another, and it is with jumps of this kind rather than with gradual changes that the new mechanics is concerned. Are these jumps final, or will they ultimately be resolved into some kind of rapid continuous motions of which we have so far no knowledge, either observational or theoretical? We simply cannot form a judgment.

The main difference between the old mechanics and the new is, however, once again a difference of background. The classical mechanics and the old quantum theory had both assumed that the whole world existed in time and space; the new mechanics is most simply expressed in terms of symbols which are best interpreted by passing beyond space and time. In transcending space and time, the new quantum mechanics finds a new background which makes for far greater simplicity and so probably comes nearer to ultimate truth. In passing from the old mechanics to the new, the mathematical description of the pattern of events stands almost unaltered, while the interpretation we put upon the symbols is utterly changed.

The history of theoretical physics is a record of the clothing of mathematical formulae which were right, or very nearly right, with physical interpretations which were often very badly wrong. When Newton had found laws of motion of a mechanical system which were true (apart from the minor refinements of the theory of relativity), he put science on a wrong track for two centuries by interpreting them in terms of forces and absolute space and time. It was much the same with his supposed force of gravitation. Again, when the true laws of the propagation of light had been discovered, they were interpreted as applying to the propagation of waves in an ether which was supposed to fill all space, and again science was started along a wrong road which it was to follow for nearly two centuries.

Now when philosophy has availed itself of the results of science, it has not been by borrowing the abstract mathematical description of the pattern of events, but by borrowing the then current pictorial description of this pattern; thus it has not appropriated certain knowledge but conjectures. These conjectures were often good enough for the man-sized world, but not, as we now know, for those ultimate processes of nature which control the happenings of the man-sized world, and bring us nearest to the true nature of reality.

One consequence of this is that the standard philosophical discussions of many problems, such as those of causality and free-will or of materialism or mentalism, are based on an interpretation of the pattern of events which is no longer tenable. The scientific basis of these older discussions has been washed away, and with their disappearance have gone all the arguments, such as they were, that seemed to require the acceptance of materialism and determinism and the renunciation of human free-will. This does not mean that the conclusions previously reached were necessarily wrong, for a bad argument may lead to a good conclusion. But it does mean that the situation must be reviewed afresh. Everything is back in the melting-pot, and we must start anew and try to discover truth on the basis of the new physics. Apart from our knowledge of the pattern of events, our tools can only be probable reasoning and the principle of simplicity.

## THE NEW PICTURE OF MODERN PHYSICS

We may appropriately start from those things of which we have the most certain knowledge, namely ourselves and our sensations. These sensations come to us through our senses, the most important of which is the sense of seeing. We see through the impact of radiation on the retina, this arriving in the form of the individual units we call photons. Other sense-organs act in a similar way, the smallest unit of sensation being produced by the arrival of a single quantum of energy from the world outside.

We have seen that photons may be represented as travelling in a space of three dimensions. This we may at once identify with the space of ordinary everyday life, because by space the ordinary man means the space through which photons travel to his eyes, the space in which he seems to see things shining or reflecting light, moving or standing still, the space in which he meets his friends.

These photons end their journeys by falling into our eyes, and so affecting our consciousness. But they are far from being projectiles falling at random. If we stand in the open on a clear night, we shall find that there are some directions of space from which photons arrive in a continuous stream and others from which no photons arrive. From such observations as this we deduce the existence of certain permanent sources of photons, or, more generally, of permanent sources of sensations; these we designate as matter.

This leads us to postulate the existence of a world of photons and matter, existing in ordinary space; it is what the plain man describes as the material world.

So far this material world has been nothing more than a mental construct private to ourselves; the space is our perceptual space, and may have no existence outside our own consciousness. If we now go asleep, or if our consciousness ceases for any other reason to function for a time, we shall find on awakening new sources of sensations which it is reasonable to identify with the old; the bedroom I find when I waken in the morning is so exactly similar to the room I left when I fell asleep that a tremendous simplicity is introduced by assuming that it is the same, and that it has been in existence all the time.

On the same principle, the moon, planets and stars outside the room may be identified with those I left behind me when I fell asleep. These, however, are no longer in the same positions. If I study these changes of position, I shall find that they are precisely those that would have occurred if the bodies had described geodesics in a curved space-time unity of the kind described on p. 63. A tremendous gain in simplicity is now secured by supposing that a curved space-time has been in existence during my sleep, and that the astronomical bodies have moved in this. Thus we conclude, with a high degree of probability, that the space-time unity and the objects which figure in it cannot be mere constructs of our individual minds, but must have existences of their own, although we know that space and time separately are abstractions of our individual minds from the space-time unity. This does not of course touch the question, to which we shall return later, of whether space, time and the material world are or are not of a mental nature, being perhaps the constructs of a consciousness superior to our own. So long as we are concerned only with our sensations, it is all the same whether we regard the world as a mental construct or as having an existence of its own independent of mind—the essential point at the moment is that it cannot be a private mental construct of our own.

## Appearance and Reality

The doctrine of materialism asserted that this space, time and material world comprised the whole of reality; it regarded consciousness as only a minor incident in the history of the material world, a somewhat exceptional episode in the haphazard muddle resulting from the chaotic movements of photons, electrons and matter in general. It interpreted thought as a mechanical motion in the brain, and emotion as a mechanical motion in the body. It seemed at one time to receive substantial support from science. For consciousness was never experienced except in conjunction with matter; a man's mental state was obviously influenced by the food, drink and drugs given to his body; and many thought it possible that all mental activities might be interpreted in terms of various physico-mental processes occurring in the associated body. At the same time astronomy was finding that only an inconceivably minute fraction of space provided any possibility for the existence of the kind of life we know, and it seemed impossible that the rest of the universe should contain anything but inanimate matter. It was hard to imagine that consciousness should be of fundamental importance in such a world.

The new physics suggests that, besides the matter and radiation which can be represented in ordinary space and time, there must be other ingredients which cannot be so represented. These are just as real as the material ingredients, but do not happen to make any direct appeal to our senses. Thus the material world as defined above constitutes the whole world of appearance, but not the whole world of reality; we may think of it as forming only a cross-section of the world of reality.

We may picture the world of reality as a deep-flowing stream; the world of appearance is its surface, below which we cannot see. Events deep down in the stream throw up bubbles and eddies on to the surface of the stream. These are the transfers of energy and radiation of our common life, which affect our senses and so activate our minds; below these lie deep waters which we can only know by inference. These bubbles and eddies show atomicity, but we know of no corresponding atomicity in the currents below.

This dualism of appearance and reality pervades the history of philosophy, again dating back to Plato. In a famous parable, Plato depicts mankind as chained in a cave in such a way that they can look only on the wall which forms the back of the cave; they cannot see the busy life outside, but only the shadows—the appearances—which objects moving in the sunshine cast on the walls of the cave. For the captives in the cave, the shadows constitute the whole world of appearance—the phenomenal world—while the world of reality lies for ever beyond their ken.

Our phenomenal world consists of the activities of matter and photons; the theatre of this activity is space and time. Thus the walls of the cave in which we are imprisoned are space and time; the shadows of reality which we see projected on the walls by the sunshine outside are the material particles which we see moving against a background of space and time, while the reality outside the cave which produces these shadows is outside space and time.

Many philosophers have regarded the world of appearance as a kind of illusion, some sort of creation or selection of our minds which had in some way less existence in its own right than the underlying world of reality. Modern physics does not confirm this view; the phenomena are seen to be just as much a part of the real world as the causes which produce them, being simply those parts of the real world which affect our senses, while the space and time in which they occur have the same sort of reality as the substratum which orders their motions. The walls of the cave and the shadows are just as real as the objects outside in the sunshine.

As the new physics has shown, all earlier systems of physics, from the Newtonian mechanics down to the old quantum theory, fell into the error of identifying appearance with reality; they confined their attention to the walls of the cave, without even being conscious of a deeper reality beyond. The new quantum theory has shown that we must probe the deeper substratum of reality before we can understand the world of appearance, even to the extent of predicting the results of experiment.

For, whatever may happen in reality, there is no reason why the shadows on the wall should change in accordance with a causal law. There will be many different arrangements of the figures out in the sunshine which all produce the same arrangement of shadows on the wall; these many arrangements will be followed by new arrangements which will not only be different in themselves but are likely to produce different shadows on the wall. It is the same with the happenings in the world of appearance; experiments that are precisely identical so far as the phenomena go may produce entirely different results. In this way causality disappears from the world of phenomena.

It comes back when we explore the substratum of reality, although in a strange new guise. Because we have only complete photons at our disposal, and these form blunt probes, the world of phenomena can never be seen clearly and distinctly, either by us or by our instruments. Instead of seeing clearly defined particles clearly located in space and executing clear-cut motions, we see only a collection of blurs—like a badly focused lantern slide. As we have seen (p. 144), this is enough of itself to prevent our ever observing strict causality in the world of phenomena.

Each blur represents the unknown entity which the particle-picture depicts as a particle, or perhaps a group of such entities. The blurs may be pictured as wave-disturbances, the intensity of the waves at any point representing the probability that, with infinitely refined probes at our disposal, we should find a particle at that point. Or again we may interpret the waves as representations of knowledge—they do not give us a picture of a particle, but of what we know as to the position and speed of motion of the particle. Now these waves of knowledge exhibit complete determinism; as they roll on, they show us know-ledge growing out of knowledge and uncertainty following un-certainty according to a strict causal law. But this tells us nothing we do not already know. If we had found new knowledge appearing, not out of previous knowledge but spontaneously and of its own accord, we should have come upon something very startling and of profound philosophical significance; actually what we find is merely what was to be expected, and the problem of causality is left much where it was.

## Mentalism or Materialism?

In addition to the dualism of appearance and reality, many pictures of the world have exhibited a second dualism, that of mind and matter or of body and soul.

This also, so far as our knowledge carries us, started with Plato. We have seen how his picture of the world consisted of forms, which exist only in our minds, and of sensible objects which, on Plato's view, display the imprint of the forms and so exemplify the qualities embodied in the forms. Plato maintained that the forms possessed a higher degree of reality than the material objects which exemplify them, so that the world was primarily a world of ideas and only secondarily a world of material objects.

We have further seen how Descartes, two thousand years later, drew a picture of the world in which mind and matter again figured, but they were now so distinct in their natures that neither could act on the other.

Then came the Idealist (or Mentalist) philosophers, who still divided the world into mind and matter, but argued that matter had no existence in its own right; it was of the same nature as mind, and existed only so far as it was a creation of mind. Under the leadership of Bishop Berkeley, they reached their conclusions by a twofold argument.

## The First Argument for Mentalism

The first was an argument we have already noticed. Galileo, Descartes, Locke and others divided the qualities of objects and substances into the two classes which Locke designated as primary and secondary. Secondary qualities are those which are perceived by the senses, and so may be differently estimated by different percipients; primary qualities are those which are essential to the object or substance and so are inherent in it whether they are perceived or not.

We have seen that physics gives no support to this division of qualities into primary and secondary. The idealists were at one with the physicists in this, but whereas the physicists consider that all physical qualities are primary, in Locke's sense of being 'utterly inseparable from the body in what state soever it be', the idealists argued that all qualities were secondary since they could be differently estimated by different percipients, a flower looking scarlet to one man but purple to another, the leg of a cheese-mite looking minute to a man but of quite a decent size to the cheese-mite, and so on. This being so, they argued, colour and size cannot be objective properties of objects; they cannot reside in the objects themselves, but in the minds perceiving the objects. And if an object is nothing but the sum of its qualities, then when all qualities reside only in percipient minds, the object itself must do the same. In brief, the object is of the nature of an idea; existence consists in being perceived by a mind.

If so, of course, an object would be non-existent when it was not being perceived by a mind. Yet the planet Pluto was certainly in existence, and impressing its image on photographic plates, many years before anyone suspected its existence. And to all appearances things go on happening inside an empty room—the fire continues to burn and the clock to keep time; when we return we find no reason for suspecting that the clock and fire have been out of existence in our absence. Berkeley got over difficulties of this kind by supposing that an object, even though it might at times not be perceived by any human mind, was yet kept permanently in existence through being continually perceived by the mind of God. Thus the whole world became an idea in the mind of God.

We have already found reasons why science can give no countenance to any arguments which suppose objects to be the sum of their secondary qualities (p. 90); they are, in brief, as follows.

Whatever capacity a red flower may have for producing a sensation of redness in a man's mind, it also has a capacity for reflecting red light whether there is anyone to see it or not, as may be very simply proved by photography. This capacity is obviously a primary quality, being 'utterly inseparable from the body in what state soever it be', and Berkeley's argument cannot touch it. Berkeley's argument fails through his not seeing that each quality such as redness must have primary ingredients as well as its alleged secondary ingredients; there is an objective scientific redness as well as the subjective philosophic redness.

### The Second Argument for Mentalism

The second line of argument ran somewhat as follows. When I hear a bell, a hammer has given a mechanical blow to a piece of metal and set it into vibration. The vibrations have been communicated in turn to the surrounding air, to my eardrums, and to a succession of elaborate pieces of mechanism and fluids inside my ears, with the result that a sequence of minute electric currents finally reaches my brain and produces certain physical changes there. These changes result in something crossing the mysterious mind-body bridge and producing certain happenings in the mind on the far side. These happenings we describe as the hearing of a bell, a purely mental idea because we might equally well experience it in a dream when there was no bell to produce it. Berkeley argued that effects must always be of the same general nature as their causes, a mechanical effect being traceable to a mechanical cause, and so on. Or, to put it rather more precisely, whatever crosses the mind-body bridge must be of the same general nature as its cause on the one side of the bridge and as its effect on the other. Thus Berkeley maintained that as the effects $A$ on the mind side of the mind-body bridge are purely mental, their causes $B$ on the body side must also be purely mental. In brief, as $A$ is an idea, and 'an idea can be like nothing but an idea', therefore $B$ also must be an idea, or of course a set of ideas.

The argument is obviously double-edged, and just as effective when reversed. For if $B$ must be of the same nature as $A$, it is equally valid to argue that $A$ must be of the same nature as $B$. Since $A$ is purely material, the argument would now prove that our mental processes must be material in their nature, as the materialists claim.

Berkeley was only able to see one side of the argument; he wished to serve theology by proving the existence of God. Before him, Descartes had been unable to see either side, claiming that mind and matter were so dissimilar, as a matter of experience, that they could have nothing in common; he too desired to serve theology—by establishing the freedom of the will. Disregarding all its theological implications, Berkeley's argument seems to provide a valid proof that mind and matter must have something in common; we can see how much real substance there is in it if we reflect on the straits to which Descartes and Leibniz were reduced when they tried to show how the opposite might be true (p. 28).

In more recent times, Bertrand Russell has expressed what is essentially the same argument in the words: 'So long as we adhere to the conventional notions of mind and matter, we are condemned to a view of perception which is miraculous. We suppose that a physical process starts from a visible object, travels to the eye, there changes into another physical process, causes yet another physical process in the optic nerve, and finally produces some effect in the brain, simultaneously with which we see the object from which the process started, the seeing being something "mental", totally different in character from the physical processes which precede and accompany it. This view is so queer that metaphysicians have invented all sorts of theories designed to substitute something less incredible....'

'Everything that we can directly observe of the physical world happens inside our heads, and consists of *mental* events in at least one sense of the word *mental*. It also consists of events which form part of the physical world. The development of this point of view will lead us to the conclusion that the distinction between mind and matter is illusory. The stuff of the world may be called physical or mental or both or neither as we please; in fact the words serve no purpose.'

If we accept this argument, the dualism of Descartes drops out of the picture altogether, and the only question left is whether we ought to say with the materialists that mind is material, or with the mentalists that matter is mental.

Whole libraries have, as Jeffreys pungently remarks, been filled with bad arguments on both sides. The materialists felt very sure, partly because of the success of science, that there was an external world of small hard atoms existing and moving in space and time, and concluded that mind must be material, and consciousness an activity of small hard atoms in space and time. The small hard atoms have now departed from science, and we picture matter as consisting mostly of empty space. Some writers have seemed to consider that this involves far-reaching philosophical consequences, and in particular, that it carries us in the direction of mentalism. It is hard to see why. Being hit by a golf-ball hurts just as much now that we know that it is little more than empty space; we realize that its material properties of solidity and hardness have not been demolished, but are merely explained in a new way.

The materialists also felt sure, partly on account of the success of science, that the absolute space and time of Newton had real existences in their own right. The physical theory of relativity now indicates—to a high degree of probability, although without absolute certainty—that space and time do not exist separately in their own right, but are subjective selections from a wider space-time unity. Some writers have argued as though this too implied a drift towards mentalism, but again it is hard to see why. Whatever degree of reality was possessed by the space and time of the older physics has not been banished from the world, but merely transferred to the space-time unity; this joint structure is every bit as objective, and may be every bit as real, as its components space and time were once thought to be separately. The two components have simply entered into a partnership, so that they now form a single entity in the eyes of the law of science, but this makes them neither less real nor more mental than before.

The physical theory of relativity has, however, other considerations to bring forward. For the materialists, space was filled with real particles, exercising on one another forces which were electric or magnetic or gravitational in their nature; these directed the motions of the particles and so were responsible for all the activity of the world. These forces were of course as real as the particles they moved.

But the physical theory of relativity has now shown (pp. 134, 171) that electric and magnetic forces are not real at all; they are mere mental constructs of our own, resulting from our rather misguided efforts to understand the motions of the particles. It is the same with the Newtonian force of gravitation, and with energy, momentum and other concepts which were introduced to help us understand the activities of the world—all prove to be mere mental constructs, and do not even pass the test of objectivity. If the materialists are pressed to say how much of the world they now claim as material, their only possible answer would seem to be: Matter itself. Thus their whole philosophy is reduced to a tautology, for obviously matter must be material. But the fact that so much of what used to be thought to possess an objective physical existence now proves to consist only of subjective mental constructs must surely be counted a pronounced step in the direction of mentalism.

The gravitational theory of relativity again brings considerations of a new kind into play. It provides an outstanding example of the truth of Einstein's general remark (p. 183) that, as experimental research advances, the fundamental laws of nature become simplified more and more, and, as in many other departments of physics, we find this simplicity residing neither in the physical facts nor in their pictorial representations, but solely in the mathematical formulae which describe the pattern of events. These seem simple to our minds because they are expressible in the kind of mathematics to which we take naturally, and studied for the pure intellectual interest we found in it before we saw it would help us to understand nature—in brief, in pure and not in applied mathematics. Thus the pure mathematician finds it much easier to interpret gravitation in terms of his science than does the mechanic or engineer. But the pure mathematician deals with the mental sphere, the mechanic and the engineer with the material. Thus the relativity theory of gravitation, because of its close association with pure mathematics, seems to carry us yet further along the road from materialism to mentalism, and the same may be said of most of the recent developments of physical science.

The new quantum theory brings still further factors into the situation. We have seen how it puts before us the two pictures which we have described as the particle-picture and the wave-picture.

The particle-picture depicts the phenomena; its ingredients are those of the ordinary picture of the material world, namely matter and radiation existing and moving in time and space.

The ingredients of the wave-picture are wave-like disturbances. Whatever a particle may be in itself, we can never experience it as a point, but if we insist on picturing it as such, then the relative intensities of the waves indicate the relative proprieties of supposing it to exist at the various points of space.

Proprieties relative to what? The answer is: Relative to our knowledge. If we know nothing about a particle except that it exists, all places are equally likely for it, so that its waves are uniformly spread throughout the whole of space. By experiment after experiment we can restrict the extent of the waves, but we can never reduce it to a point, or indeed below a certain minimum; the coarse-grainedness of our probes precludes this, so that there must always be a finite region of wave-disturbance left. The waves in this region depict our knowledge and its imperfections exactly and precisely.

Thus the ingredients of the particle-picture are particles existing and moving in physical space, while the ingredients of the wave-picture are mental constructs existing and moving in conceptual spaces; the ingredients of the particle-picture are material, those of the wave-picture mental.

The first complete particle-picture was provided by Newton's mechanics in conjunction with his corpuscular theory of light. The mechanics supposed that those permanent sources of sensation which we call matter consisted of particles moving in physical space, while the corpuscular theory of light further supposed that the radiation by which our sense-organs are affected also consist of particles. This scheme was found not to give an adequate account of the facts of observation, and in due course the corpuscular picture of light was replaced by the present wave-picture. This resulted in complete agreement with the facts of observation so far as optical phenomena were concerned. But, until the theory of relativity appeared, it was not suspected that the ingredients of this picture were purely mental constructs.

Thus physics continued to believe that it was studying an objective nature which existed in its own right independently of the mind which perceived it, and had existed from all eternity whether it was perceived or not; this belief was the soil in which materialism had its roots. Physics would have gone on holding this belief to this day, had the electron which the physicist observed behaved as, on this supposition, it ought to have done.

But it did not so behave, and the new quantum theory was brought into existence to make good the defects. It discovered what we believe to be the true pattern of events, with the wave-picture of matter as its pictorial representation. The particle-picture of radiation had already given place to a wave-picture; it now appeared that the particle-picture of matter must also be replaced by a wave-picture. The result was a complete agreement with experiment. In this progress towards the truth, let us notice that each step was from particles to waves, or from the material to the mental; the final picture consists wholly of waves, and its ingredients are wholly mental constructs.

We must remember that this picture is not a picture of reality, it is a picture we draw to help us imagine the course of events in reality. Thus we are not entitled to argue that reality is like the ingredients of the picture, although there is a certain presumption that the two are not altogether dissimilar in their natures; the pictorial representation does not take us into the mansion of reality, but does take us to its doorstep. Thus, when it was believed that the course of events could be most easily understood in terms of forces and mechanical models, most people thought that the picture or model must be like the reality, and jumped to the conclusion that reality was mechanical in its nature. Before this, when the course of events had seemed to be governed by the caprices and passions of gods and demons, it had been assumed that reality was of a similar nature; we have seen how Thales maintained that all things must be full of gods. And now that we find that we can best understand the course of events in terms of waves of knowledge, there is a certain presumption—although certainly no proof—that reality and knowledge are similar in their natures, or, in other words, that reality is wholly mental.

Apart from arguments of this type, we can have no means of knowing the true nature of reality. The most we can say is that the cumulative evidence of various pieces of probable reasoning makes it seem more and more likely that reality is better described as mental than as material.

Even if the two entities which we have hitherto described as mind and matter are of the same general nature, there remains the question as to which is the more fundamental of the two. Is mind only a by-product of matter, as the materialists claimed? Or is it, as Berkeley claimed, the creator and controller of matter?

Before the latter alternative can be seriously considered, some answer must be found to the problem of how objects can continue to exist when they are not being perceived in any human mind. There must, as Berkeley says, be 'some other mind in which they exist'. Some will wish to describe this, with Berkeley, as the mind of God; others with Hegel as a universal or Absolute mind in which all our individual minds are comprised. The new quantum mechanics may perhaps give a hint, although nothing more than a hint, as to how this can be.

In the particle-picture, which depicts the phenomenal world, each particle and each photon is a distinct individual going its own way. When we pass one stage further towards reality we come to the wave-picture. Photons are no longer independent individuals, but members of a single organization or whole—a beam of light—in which their separate individualities are merged, not merely in the superficial sense in which an individual is lost in a crowd, but rather as a raindrop is lost in the sea. The same is true of electrons; in the wave-picture these lose their separate individualities and become simply fractions of a continuous current of electricity. In each case, space and time are inhabited by distinct individuals, but when we pass beyond space and time, from the world of phenomena towards reality, individuality is replaced by community.

It seems at least conceivable that what is true of perceived objects may also be true of perceiving minds; just as there are wave-pictures for light and electricity, so there may be a corresponding picture for consciousness. When we view ourselves in space and time, our consciousnesses are obviously the separate individuals of a particle-picture, but when we pass beyond space and time, they may perhaps form ingredients of a single continuous stream of life. As it is with light and electricity, so it may be with life; the phenomena may be individuals carrying on separate existences in space and time, while in the deeper reality beyond space and time we may all be members of one body. In brief, modern physics is not altogether antagonistic to an objective idealism like that of Hegel.

The new dualism of the particle- and wave-pictures is in many ways reminiscent of the old dualism of Descartes. There is no longer a dualism of mind and matter, but of waves and particles; these seem to be the direct, although almost unrecognizable, descendants of the older mind and matter, the waves replacing mind and the particles matter. The two members of this dualism are no longer antagonistic or mutually exclusive; rather they are complementary. We need no longer devise elaborate mechanisms, as Descartes and Leibniz did, to keep the two in step, for one controls the other—the waves control the particles, or in the old terminology the mental controls the material.

## THE PROBLEM OF FREE-WILL

We have seen how the materialists interpreted thought and emotion as mechanical activities of the brain and body respectively, and imagined that if all the physical and chemical changes in a brain and body could be traced out, it would be possible, at least in principle, to deduce all the mental and emotional experiences of the associated mind. Thus, if material changes were bound by a causal chain, mental and emotional experiences would also be so bound, and there could be no room left for free-will.

There were nevertheless two schools of thought—the *determinists* who maintained that all events, including human acts, were causally determined and so compelled by past events and acts, including such events as those of heredity, environment, acquired habits and so forth; and the *indeterminists* who maintained that human acts are not entirely determined by the past, but that at every moment we can exercise a certain amount of guidance through a fiat which is our own.

On the determinist view, a man's actions would of course be completely predictable in principle by one who had a sufficiently intimate knowledge of his nature, of his past and of the character he has acquired in the past. On the indeterminist view, this is not so; a man can falsify all predictions by a capricious, and so unpredictable, choice.

### The Determinists

Practically all modern philosophers of the first rank—Descartes, Spinoza, Leibniz, Locke, Hume, Kant, Hegel, Mill, Alexander, as well as many others—have been determinists in the sense of admitting the cogency of the arguments for determinism, but many have at the same time been indeterminists in the sense of hoping to find a loophole of escape from these arguments. Often they conceded that our apparent freedom is an illusion, so that the only loophole they could hope to find would be an explanation as to how the illusion could originate.

Descartes and Kant, as we have seen, may fairly be described as determinists trying to shed their determinism, while Leibniz, Locke and Hume are perhaps better described as determinists trying to explain their determinism. Spinoza, Mill and Alexander were out-and-out determinists, although like many other determinists they were not always consistent in their determinism.

Leibniz thought that there are always sufficient reasons in the nature and character of each one of us to determine for us any decision we may be called upon to make. We are, then, never free, because our acts at every moment are completely determined by our nature which came to us in the past, and by our character which was formed in the past. Hume also thought that our decisions are always determined by our characters, so that to make a different decision we should need to be a different person. Locke thought our decisions are based on our desires to enjoy pleasure and avoid pain, and so are determined by our estimates of future pleasure and pain—although of course our judgments may be wrong. Spinoza thought that our actions and experiences are in actual fact determined by a sort of mathematical necessity, like that of a wheel in a machine, but that we feel ourselves free if we enjoy doing what actually we are doing under compulsion; a stone in the air, he said, would think itself free if it could forget the hand that had thrown it. Or, to take a more homely illustration which is not Spinoza's, I know that I choose jam-roll because I like it, and I feel myself free in so choosing because I do not stop to think that my liking is the inevitable result of my inheritance and upbringing, of the present state of my health and of my sugar metabolism, and of all sorts of things which it is quite beyond my power to change at the moment. Hegel and, at a later period, Alexander held very similar opinions. Kant thought that we feel ourselves free just in so far as our actions appear rational to us; if I rationally run downstairs to welcome a friend, my action seems free to me, but if I run downstairs irrationally because I am afraid of a ghost, it will seem to me that I acted under compulsion. Mill believed that all human actions are so completely determinate that sociology could be made into a perfectly exact science, in which the future of a society would be seen to follow from its past with a mechanical certainty and after unvarying laws. He then, with the characteristic irrationality of the thoroughgoing determinist, wanted these laws to be studied with a view to improving the future of the race!

The average plain man who is no philosopher will probably consider that the springs of human action are too varied, too intricate and too complex to be summed up in any single formula. His own philosophy is not very clear-cut, but may perhaps be described as one of determinism for others and freedom for himself. Yet this supposed freedom applies only to his present acts, and not to the past; we see our past selves as other men. For, as Henry Sidgwick says: 'We always explain the voluntary action of all men except ourselves on the principle of causation by character and circumstances. We infer generally the future actions of those whom we know from their past actions; and if our forecast turns out in any case to be erroneous, we do not attribute the discrepancy to the disturbing influence of free-will, but to our incomplete acquaintance with their character and motives.... Nay even as regards our own actions, however free we feel ourselves at any moment, however unconstrained by present motives and circumstances and unfettered by the result of what we have previously been and felt our volitional choice may appear, still when it is once well past, and we survey it in the series of our actions, its relations of causation and resemblance to other parts of our life appear, and we naturally explain it as an effect of our nature, education and circumstances.'

Not only so, but the freedom we claim for our present selves is almost indistinguishable from the determinism we attribute to others. We usually claim no freedom for ourselves beyond that of being able to do what we want to do, which simply means yielding to the strongest impulse, the freedom of the beam of the weighing scale to incline to the heavier side, the kind of freedom which philosopher and scientist agree in describing as determinism—since, under it, the future is fully determined; it follows from the past with the inevitability of a machine.

We can see this by examining special instances. Mr Average Man thinks over his past, and proclaims that if he were young again, he would choose a different profession. He may insist that he would be free to make his own choice, but all he means is that if, at the age of eighteen, he had had the knowledge and experience of life which he now has at fifty, he would have acted differently. Of course he would, and so would we all, but this is no evidence of freedom. If Mr Man now had to make his choice again, with precisely the same knowledge and experience as he had at eighteen, he would review the situation in the same way as he did before, the same considerations would be thrown into the scales, and the balance would again swing in the same direction as before. He will not claim a freedom to act from pure caprice, but only a freedom to yield to the strongest motive—the freedom of Newton's apple to fall towards the earth rather than towards the moon, because the earth attracted it more forcibly than the moon. And this is not freedom of any kind; it is pure determinism. As Hume said, to have made a different decision, he would have had to be a different man.

Or perhaps he may claim he is free to choose in trivial matters, as for instance whether he will ask for black or white coffee. Perhaps he usually asks for black, and if on some rare occasion he asks for white, he may imagine that in so trivial a matter his choice was wholly undetermined. But a psychologist will tell him that, even here, he can only yield to the strongest motive, no matter how weak these motives may be. When he made his unusual choice, his mind may have been far away from food and drink, absorbed in the pages of a book he was looking at, so that, when politeness compelled him to make a choice, he merely mentioned the colour suggested by the pages of his book. Or he may have felt a temporary but unconscious aversion to black and blackness through some association, such as mourning or a funeral. There are endless possibilities and only one impossibility, which is that he said 'white' out of pure caprice, without having any guiding motive in his mind. The presence of milk in his coffee in two minutes' time will be a direct consequence of the state of his mind now, just as surely as the state of the material universe in two minutes' time will, on the deterministic view, be a direct consequence of its state now.

Although Mr Average Man may occasionally protest that he is incapable of acting meanly or dishonourably, yet in general he would hate to think that he is not free to choose his own course of action at every moment of his life. Thus he likes to think that his own actions are wholly unpredictable, and yet, when other men behave in a wholly unpredictable way, he describes them as weak fools. In brief, freedom in ourselves is a virtue, but in others a vice; freedom is something we possess, but that others do not.

Not only plain men, but philosophical writers also, seem to confuse free-will with determinism of this unconscious kind. Thus Henry Sidgwick (*Methods of Ethics*) says that the question at issue in the free-will controversy, as he understands it, is whether his action at any moment is completely determined by his character and the external influences, including his bodily condition, which act on him at the moment, 'or is there always a possibility of my choosing to act in the manner that I now judge to be reasonable and right, whatever my previous actions and experiences may have been?'

But a judgment as to what is reasonable and right cannot be based on nothing at all—if it is, it is no judgment but pure caprice. And it cannot be based on anything other than a man's character, which is founded in his previous actions and experiences, and the external influences acting on him at the moment—in brief, on the past and the present, or on what is inside him and on what is outside him. Thus Sidgwick's second alternative, which is clearly intended to represent free-will, is that our actions are determined by our judgments, and our judgments by our inner character and external influences—which brings us round to precisely his description of determinism. Thus his two alternatives are not determinism and freedom at all, but merely conscious and unconscious determinism, and he never reaches the real issue of free-will.

The same is true of theological attempts to solve the problem by adding Divine intervention to the external influences acting on a man—'We have no power to do good works...without the grace of God by Christ preventing us, that we may have a good will, and working with us when we have that good will.' Such Divine intervention does not add to a man's freedom, but to the restrictions on it.

Attempts have been made to find an alternative to determinism in what is described as 'teleological causation', according to which the future determines the present, or at least influences it, like the legendary carrot held in front of the donkey's nose. If a student is working hard in the hope of passing an examination, it is argued that the present spell of hard work is the effect of a future cause, namely an examination which is to be held at some future date. But it is surely more true to say that the cause is not the examination—which after all may never take place, and so can hardly be the cause of something which has already taken place—but the hope of passing the examination. This hope is not in the future; a man will not be working for an examination at this moment unless the hope of passing it has been in his mind at some previous moment, so that the proximate cause of his hard work is in the past, and not in the future. To some extent, the whole matter is one of juggling with words, but in whatever sense words are used, ideas such as teleological causation can throw no new light on the ultimate problem.

## The Indeterminists

On the other side, Lotze (1817–1881) and William James (1842–1910) were consistent and logical indeterminists. Lotze agreed with the determinists that both natural events and human acts lie on strings of causal chains, and that such causal chains when once started have no end in future time, but he thought that such chains may have capricious beginnings. William James advocated the doctrine he described as tychism—chance playing its part in ordering the course of events. According to him, the pattern of events is not unalterably fixed; we introduce novelties when we make choices (but it is not explained why one novelty rather than another is introduced).

We have already seen that modern physics is not entirely hostile to such ideas in their application to inanimate nature, although we also saw (p. 178) that they should not be applied to the underlying realities, but only to the phenomena as seen and understood by us—in other words, the indeterminacy does not reside in objective nature, but only in our subjective interpretation of nature.

Let us, however, ignore the distinction, and state the case in the form most favourable to indeterminism and freedom by imagining that an assigned state $A$ of the inanimate world may be followed by any one of a number of different states $B, C, D, \ldots$ all of which lead to different future states of the world. In the inanimate world we find no apparent reason why $A$ should be followed by $B$ rather than by $C$ or $D$. But suppose that in situations in which the human mind is concerned, the mind has some power of directing minute bits of the world to any one of the states, $B, C, D, \ldots$ as it chooses. Since all the transitions $A \to B$, $A \to C$, $A \to D$, etc. conform to the conservation of energy and momentum, we have mind acting on matter without the exercise of any material force or any transfer of energy, and moulding the universe within limits to its choice. This brings us to something very like Descartes's original explanation of the action of mind on matter (p. 25), but it is no longer open to the objections raised by Leibniz.

Essentially the same solution was propounded by Clerk Maxwell. The course of a railway train is uniquely prescribed for it at most points of its journey by the rails on which it runs. Here and there, however, it comes to a junction at which alternative courses are open to it, and it may be turned on to one or the other by the quite negligible expenditure of energy involved in moving the points. Maxwell thought that the human body might come to similar junctions, at which it could be turned into one course or another by the action of the mind, without any expenditure of mechanical energy—the body is the train, the mind is the pointsman. The indeterminacy of atomic motions has seemed to many to provide just the kind of junction, and possibly also of points, that Maxwell needed.

This may suggest a possible way in which mind can act on matter, but it leaves the deeper problem of freedom of choice untouched. Even if the pointsman can move the points and divert the motion of the train in so doing, the question of why he moves the points in one direction rather than in another remains. If he moves them according to a pre-arranged plan, the train is simply following a schedule, which makes its motion as determinate as if the points and junction were non-existent. If, as most people would say, he moves them in a particular direction 'because he chooses to', the question is why he chooses this direction rather than the other. If something determines his choice, we are back to determinism; if nothing, he acts from pure caprice, and this leads to a free-will which is neither of the kind we want to find nor of the kind we feel we do find. We like to imagine that we hold determinism at bay by our wisdom or virtue or foresight, and not through a mere random caprice over which we have no control and so for which we are in no way responsible. A man who has done a foolish deed may find comfort in thinking that he was the plaything of capricious forces, but not so the man who has been prudent or generous or has put his money on a winner.

Neither does a capricious indeterminism give us a free-will at all resembling that of our experience or imagined experience. If every event were not determined by a sufficient reason, the whole world would, as Leibniz remarked, be a chaos. A mind endowed with free-will of the capricious variety would be a prey to spontaneous and wholly irrational impulses; we should describe it as the mind of a madman, although in actual fact no madman's mind is ever quite so crazy. The further psychology and common sense probe into the question, the more necessary they find it to accept orthodox determinism—our acts are determined by our volitions, our volitions by our motives, and our motives by our past. The psychologist will think of this past in terms of heredity and environment, the moralist in terms of ethical and spiritual influences, and the physiologist in terms of physico-chemical activities. But all will agree that the relative strength of the various motives is determined by past events, so that a man never chooses for himself; his past always chooses for him.

### Present-day Opinion

Notwithstanding the apparent want of determinism disclosed in inanimate nature by the quantum theory, this is still the opinion of the vast majority of present-day physicists. Thus in his book, *Where is science going?* Planck, the founder of the quantum theory, writes: 'No biographer will attempt to solve the question of the motives that govern the acts of his hero by attributing these to mere chance. He will rather attribute his inability to the lack of source materials, or he will admit that his own powers of spiritual penetration are not capable of reaching down into the depths of these motives. And in practical everyday life our attitude to our fellow-beings is based on the assumption that their words and actions are determined by distinct causes, which lie in the individual nature itself or in the environment, even though we admit that the source of these causes cannot be discovered by ourselves.... The principle of causality must be held to extend even to the highest achievements of the human soul. We must admit that the mind of each one of our great geniuses—Aristotle, Kant or Leonardo, Goethe or Beethoven, Dante or Shakespeare—even at the moments of its highest flights of thought or in the most profound inner workings of his soul—was subject to the causal fiat and was an instrument in the hands of an almighty law which governs the world.'

In the same book, Einstein is reported as saying: 'Honestly I cannot understand what people mean when they talk about the freedom of the will. I feel that I will to light my pipe and I do it, but how can I connect this up with the idea of freedom? What is behind the act of willing to light the pipe? Another act of willing? Schopenhauer once said: Der Mensch kann was er will; er kann aber nicht wollen was er will.'

Modern philosophy also seems to have come to the conclusion that there is no real alternative to determinism, with the result that the question now discussed is no longer whether we are free but why we think we are free. We have seen how Alexander divides the world into levels which are at different stages of evolution— space-time, matter, life, mind, Deity. While conceding that all events are in actual fact deterministic, he considers that the inhabitants of each level may feel themselves free, while noting the absence of freedom prevailing in the levels lower than their own. Thus atoms, in the lowest level but one, feel themselves free when they contemplate space-time in which no freedom is possible; we have already quoted Spinoza's remark that a stone in the air would think itself free if it could forget the hand that had thrown it. In the same way, we think ourselves free, but think that machines and even plants—the levels just beneath us—are determinate. And again God, contemplating our activities from His higher level, feels Himself free but sees that we are not.

Without accepting any such scheme in detail, many philosophers would agree that we are able to do what we wish within limits, and so feel ourselves free, but this is only because we do not pause to reflect that our wishes themselves—the springs of our actions—are thrust on us by our pasts. On the other hand, as we have no immediate experience of this feeling of freedom in others, we see that their acts are thrust on them by their pasts, and so regard these acts as determinate.

In brief, neither the philosophical study nor the physical research of the last 300 years has shown any cause for changing Descartes's dicta that 'nothing cannot be the efficient cause of anything' and that 'the power of the will consists only in this, that...we so act that we are not conscious of being determined to a particular action by any external force'. Thus free-will is only our name for unconscious determinism. But Kant would presumably have argued that all this does not prove that we are devoid of freedom, so much as that a deterministic way of looking at things is ingrained in our minds; it is our way of interpreting the temporal sequence of events.

And of course it may be. After a few individual experiences of the type 'I have bumped my head, and I feel a pain', the growing child generalizes to such propositions as 'I have bumped my head, and *therefore* I feel a pain' and 'If I bump my head, I shall feel a pain'. Such associations of ideas prove helpful in avoiding further misadventures, and so are extended, and the habit of finding cause-effect relations grows. But there is a continuous transition from cases such as those just mentioned to 'It is night, so it will soon be day' or 'I am hungry, so shall soon get something to eat', which are not cause-effect relations at all. In these and similar ways the *post hoc ergo propter hoc* habit of mind may become ingrained, and it may be possible to find a perfectly simple psychological explanation of the cause-effect habit of the human mind without even calling upon any inborn mental 'category'.

In any case there can be no question that all our conscious

experiences of inanimate nature, which are limited to the man-sized world, show that determinism does prevail here. It may be that, because of this, we are unable to imagine how anything but determinism can govern the inanimate world—although modern physics shows that it does so far as the phenomena are concerned—and that we then transfer this inhibition from the material to the mental world. If so, it is neither abstract physics nor concrete experience that thrusts determinism upon us, but rather the inability of our minds to imagine anything other than determinism.

Before the era of modern physics, it was a simple matter to define what we meant by causality and free-will. We supposed the world to consist of atoms and radiation; we imagined that precise positions could be assigned, in principle, to every atom and to every element of radiation, and the question of causality was simply whether, knowing these positions, it was possible in principle to predict the future course of events with certainty. The question of free-will was whether it was still possible to predict this course when consciousness and human volitions intervened in the picture.

But modern physics shows that these formulations of the questions have become meaningless. It is no longer possible to know the exact positions of particles or of elements of radiation, and, even if we could, it would still be impossible to predict what was going to happen next. So far as the inanimate world is concerned, we may picture a substratum below space and time in which the springs of events are concealed, and it may be that the future already lies hidden, but uniquely and inevitably determined, in this substratum. Such a hypothesis at least fits all the known facts of physics. But as we pass from the phenomenal world of space and time to this substratum, we seem, in some way we do not understand, to be passing from materialism to mentalism, and so possibly also from matter to mind. It may be then that the springs of events in this substratum include our own mental activities, so that the future course of events may depend in part on these mental activities.

At least the new physics has shown that the problems of causality and free-will are in need of a new formulation. If those who believe in freedom of the will could explain what they mean by freedom, and could show precisely where it differs from what we have called unconscious determinism, it is at least conceivable that what they want would be found in modern physics. The classical physics seemed to bolt and bar the door leading to any sort of freedom of the will; the new physics hardly does this; it almost seems to suggest that the door may be unlocked—if only we could find the handle. The old physics showed us a universe which looked more like a prison than a dwelling-place. The new physics shows us a universe which looks as though it might conceivably form a suitable dwelling-place for free men, and not a mere shelter for brutes—a home in which it may at least be possible for us to mould events to our desires and live lives of endeavour and achievement.

## CONCLUSION

There is a temptation to try to round off our discussion by summarizing the conclusions we have reached. But the plain fact is that there are no conclusions. If we must state a conclusion, it would be that many of the former conclusions of nineteenth-century science on philosophical questions are once again in the melting-pot.

Just because of this, we cannot state any positive conclusions of any kind, as for instance that materialism is dead, or that a deterministic interpretation of the world is obsolete, but we can say that determinism and freedom, matter and materialism need to be redefined in the light of our new scientific knowledge. When this has been done, the materialist must decide for himself whether the only kind of materialism which science now permits can be suitably labelled materialism, and whether the ghostly remains of matter should be labelled as matter or as something else; it is mainly a question of terminology.

What remains is in any case very different from the full-blooded matter and the forbidding materialism of the Victorian scientist. His objective and material universe is proved to consist of little more than constructs of our own minds. In this and in other ways, modern physics has moved in the direction of mentalism.

Again we can hardly say that the new physics justifies any new conclusions on determinism, causality or free-will, but we can say that the argument for determinism is in some respects less compelling than it seemed to be fifty years ago. There appears to be a case for reopening the whole question as soon as anyone can discover how to do so.

This may seem a disappointing harvest to have garnered from so extensive a field of new scientific activity, and from one, moreover, which comes so close to the territory of philosophy. Yet we may reflect that physics and philosophy are at most a few thousand years old, but probably have lives of thousands of millions of years stretching away in front of them. They are only just beginning to get under way, and we are still, in Newton's words, like children playing with pebbles on the sea-shore, while the great ocean of truth rolls, unexplored, beyond our reach. It can hardly be a matter for surprise that our race has not succeeded in solving any large part of its most difficult problems in the first millionth part of its existence. Perhaps life would be a duller affair if it had, for to many it is not knowledge but the quest for knowledge that gives the greater interest to thought—to travel hopefully is better than to arrive.

# INDEX